Architectural Interior Lighting

Gurkan Ozenen

Architectural Interior Lighting

Springer

Gurkan Ozenen
Istanbul, Türkiye

ISBN 978-3-031-49694-3 ISBN 978-3-031-49695-0 (eBook)
https://doi.org/10.1007/978-3-031-49695-0

This Springer imprint is published by the registered company Springer Nature Switzerland AG
The registered company address is: Gewerbestrasse 11, 6330 Cham, Switzerland

Paper in this product is recyclable.

I dedicate this book to my cherished daughter and my lovely wife. Their unwavering love, support, and encouragement have been the driving force behind my pursuit of knowledge and passion for lighting design. I would like to thank them for always inspiring me to reach for the stars and for being my constant source of strength and joy. This book is a tribute to my beautiful family and the love that binds us together.

Preface

Lighting design is an essential component of architectural design that can have a significant impact on the aesthetic and functional aspects of a space. As an architect, I have come to appreciate the importance of lighting in creating a comfortable and visually appealing environment. The focus of this book, *Architectural Interior Lighting*, is to provide an in-depth understanding of lighting principles and practices that are necessary for the creation of visually appealing and functional interiors.

The book is divided into nine chapters, each of which covers a particular aspect of interior lighting design. Chapter 1 is a review of the terminology used in lighting design. Chapter 2 provides an overview of light and color theory, including the physiological and psychological effects of lighting. Chapter 3 focuses on lighting fundamentals and design principles, including lighting metrics, lighting layers, and the lighting triangle.

Chapter 4 discusses the basic principles of lighting, including the types of light sources, their properties, and their applications in interior design. Chapter 5 covers lighting sources used for interior design, including natural light, artificial light, and their combinations. Chapter 6 delves into decorative lighting for interior design, including chandeliers, pendants, and other decorative fixtures.

Chapter 7 discusses professional lighting for interior design, including task lighting, accent lighting, and specialty lighting. Chapter 8 focuses on lighting controls and systems, including dimming and scene control systems. Finally, Chap. 9 explores sustainable lighting design, including energy-efficient lighting sources, lighting controls, and daylighting.

Throughout the book, I have tried to present the subject matter in an accessible and informative manner, using examples and illustrations to enhance understanding. My hope is that this book will provide architects, interior designers, and lighting professionals with the knowledge and skills necessary to design beautiful and functional interior spaces.

I am grateful to Springer for the opportunity to share my passion for lighting design and to contribute to the field of interior design. I would also like to thank my family for their support and encouragement throughout this project.

Istanbul, Turkey Gurkan Ozenen

Summary

Architectural Interior Lighting is a comprehensive guide that covers various aspects of lighting design in interior spaces. This book aims to provide architects, interior designers, and lighting designers with a comprehensive understanding of the principles and techniques of lighting design for various interior spaces.

The book discusses different types of lighting sources, including artificial and natural lighting, and their respective applications in interior design. It also covers various lighting techniques such as accent, task, decorative, and ambient lighting and provides practical examples of their implementation in different interior spaces.

The book delves into professional lighting features such as high-quality lighting fixtures, layered lighting, lighting controls, and color temperature control, explaining how these elements work together to create a harmonious and functional lighting design.

Ultimately, this book serves as a valuable resource for both professionals and students interested in the field of architectural interior lighting. It contributes to research by providing a comprehensive overview of lighting design strategies, and it offers practical insights that can be applied in real-world design projects.

This book provides a comprehensive overview of architectural interior lighting, including various lighting techniques, fixtures, controls, and calculations, to enhance the functionality and aesthetics of interior spaces.

What's unique, different, and engaging about this book is that this book is a comprehensive guide to understanding the role of lighting in interior design. This book provides readers with an in-depth understanding of how lighting can be used to enhance the functionality, aesthetics, and ambiance of an interior space.

Unlike other books on interior lighting, this guide goes beyond simply providing basic lighting principles and techniques. It delves into the advanced concepts and strategies used by professionals to create stunning and impactful lighting designs.

This book is perfect for interior designers, architects, lighting designers, and anyone interested in creating beautiful and functional interior spaces. It is a must-read for professionals who want to take their lighting design skills to the next level. With detailed descriptions, this book is an engaging and informative resource that readers will return to again and again. Whether you are looking to design a residential space, commercial space, or public building. This book provides the tools and knowledge needed to create stunning lighting designs that elevate the ambiance and functionality of any interior space.

Contents

Author's Biography

Gurkan Ozenen is an accomplished academician and architect with an impressive educational background. He earned his Bachelor of Science (B.S.) degree in architecture with honors from Mimar Sinan Fine Art University in 2004. Following this, he completed his Master's (M.Sc.) degree and Doctor of Philosophy (Ph.D.) degree in Architectural Design Computing at Istanbul Technical University in 2007 and 2016, respectively.

Dr. Ozenen began his academic career as a full-time lecturer with the rank of Assistant Professor at Dogus University in 2017, where he served until 2021. He then joined Istanbul Health and Technology University as a full-time university lecturer with the rank of Assistant Professor in 2021, where he continues to work.

As an expert in architectural design computing, Dr. Ozenen has conducted extensive research in the fields of generative design, digital fabrication, and computational design. His research interests also include building information modeling, parametric design, and interactive architecture.

Dr. Ozenen has also taught various courses related to architectural interior design, architectural interior lighting, architectural design computing, including digital fabrication, design algorithms, and interactive architecture. Dr. Ozenen's dedication to academia and his extensive research in the field of architecture make him a valuable asset to the academic community.

Chapter 1
Terms Used in Lighting

Contents

Some common terms used in lighting, along with their academic references are as follows:

- Illuminance: Illuminance is the amount of light falling on a surface per unit area. It is usually measured in lux (lx). The standard reference for illuminance is CIE 69-1987, "Method of Measuring and Specifying the Photometric Characteristics of Light Sources."
- Luminance: Luminance is the amount of light reflected off a surface per unit area in a given direction. It is usually measured in candelas per square meter (cd/m^2). The standard reference for luminance is CIE 15.2-1986, "Colorimetry."
- Color temperature: Color temperature is a measure of the color of light emitted by a source. It is usually measured in Kelvin (K). The standard reference for color temperature is CIE 15.2-1986, "Colorimetry."
- Color rendering index (CRI): CRI is a measure of how well a light source renders colors compared to a reference source. It is usually measured on a scale from 0 to 100. The standard reference for CRI is CIE 13.3-1995, "Method of Measuring and Specifying Colour Rendering Properties of Light Sources."
- Spectral power distribution (SPD): SPD is a measure of the distribution of energy across the spectrum of a light source. It is usually measured in watts per nanometer (W/nm). The standard reference for SPD is CIE 15.2-1986, "Colorimetry."

G. Ozenen, *Architectural Interior Lighting*,
https://doi.org/10.1007/978-3-031-49695-0_1

- Flicker: Flicker is the rapid fluctuation of light output from a source. It is usually measured in percentage or frequency (Hz). The standard reference for flicker is IEC 61000-3-3:2013, "Electromagnetic compatibility (EMC) – Part 3-3: Limits – Limitation of voltage changes, voltage fluctuations and flicker in public low-voltage supply systems, for equipment with rated current ≤16 A per phase and not subject to conditional connection."
- Glare: Glare is the visual discomfort caused by excessive brightness or contrast. It is usually measured in terms of luminance ratio or visual comfort probability (VCP). The standard reference for glare is CIE 117-1995, "Discomfort Glare in Interior Lighting."

1.1 Illuminance

Illuminance is the measure of the amount of light that falls on a surface per unit area, usually measured in lux (lx). Illuminance is an important factor in lighting design, as it determines the brightness of a space and can affect visual comfort, performance, and safety.

The standard reference for illuminance is CIE 69-1987, "Method of Measuring and Specifying the Photometric Characteristics of Light Sources." This document provides guidelines for measuring illuminance using various instruments and techniques, and it also includes recommendations for specifying illuminance levels for different applications [1].

Illuminance is influenced by a number of factors, including the intensity and distribution of the light source, the distance between the source and the surface, and the reflectance and orientation of the surface. In general, higher illuminance levels are needed for tasks that require greater visual acuity, such as reading or precision work, while lower levels may be sufficient for general ambience or relaxation.

However, it is important to note that illuminance alone does not fully determine the visual quality of a space. Other factors, such as luminance, color temperature, and color rendering, can also affect visual perception and comfort.

In practice, illuminance is often used as a basis for lighting design and assessment, and various standards and guidelines exist to provide recommendations for illuminance levels in different applications. For example, the Illuminating Engineering Society (IES) publishes the Lighting Handbook, which includes illuminance recommendations for various indoor and outdoor spaces based on factors such as occupancy, task, and visual needs.

In summary, illuminance is an important measure of light in lighting design and assessment. It is influenced by a number of factors and can affect visual comfort, performance, and safety. References such as CIE 69-1987 and the IES Lighting Handbook provide guidelines and recommendations for measuring and specifying illuminance levels in different applications [1].

1.2 Luminance

Luminance is the measure of the amount of light that is reflected off a surface per unit area in a given direction, usually measured in candelas per square meter (cd/m^2). Luminance is an important factor in lighting design, as it determines the perceived brightness and contrast of a space and can affect visual comfort, performance, and safety.

The standard reference for luminance is CIE 15.2-1986, "Colorimetry." This document provides guidelines for measuring luminance using various instruments and techniques, and it also includes recommendations for specifying luminance levels for different applications.

Luminance is influenced by a number of factors, including the amount and quality of the light source, the reflectance and texture of the surface, and the viewing angle and distance. In general, higher luminance levels are needed for tasks that require greater visual acuity, such as reading or precision work, while lower levels may be sufficient for general ambience or relaxation [2].

However, it is important to note that luminance alone does not fully determine the visual quality of a space. Other factors, such as illuminance, color temperature, and color rendering, can also affect visual perception and comfort.

In practice, luminance is often used as a basis for lighting design and assessment, and various standards and guidelines exist to provide recommendations for luminance levels in different applications. For example, the Illuminating Engineering Society (IES) publishes the Lighting Handbook, which includes luminance recommendations for various indoor and outdoor spaces based on factors such as occupancy, task, and visual needs.

In summary, luminance is an important measure of light in lighting design and assessment. It is influenced by a number of factors and can affect visual comfort, performance, and safety. References such as CIE 15.2-1986 and the IES Lighting Handbook provide guidelines and recommendations for measuring and specifying luminance levels in different applications [2].

1.3 Color Temperature

Color temperature is a measure of the color appearance of a light source, usually expressed in degrees Kelvin (K). Color temperature is an important factor in lighting design, as it affects the perceived color and mood of a space and can also impact visual performance and comfort.

The standard reference for color temperature is CIE 15:2004, "Colorimetry." This document provides guidelines for measuring and specifying color temperature using various instruments and techniques, and it also includes recommendations for color temperature values for different lighting applications.

Color temperature is influenced by the spectral distribution of the light source, which determines the balance of colors in the light. Lower color temperatures, around 2000–3000 K, are considered “warm” and have a yellowish-red hue, while higher color temperatures, around 5000–6500 K, are considered “cool” and have a bluish-white hue. Neutral white light has a color temperature of around 4000 K.

In practice, color temperature is often used as a basis for lighting design and assessment, and various standards and guidelines exist to provide recommendations for color temperature values in different applications. For example, the Illuminating Engineering Society (IES) publishes the Lighting Handbook, which includes color temperature recommendations for various indoor and outdoor spaces based on factors such as occupancy, task, and visual needs.

However, it is important to note that color temperature alone does not fully determine the perceived color quality of a light source. Other factors, such as color rendering, can also affect the appearance of colors under the light.

In summary, color temperature is an important measure of the color appearance of a light source in lighting design and assessment. It is influenced by the spectral distribution of the light source and can affect the perceived color and mood of a space as well as visual performance and comfort. References such as CIE 15:2004 and the IES Lighting Handbook provide guidelines and recommendations for measuring and specifying color temperature values in different applications.

1.4 Color Rendering Index (CRI)

Color rendering index (CRI) is a measure of a light source’s ability to accurately render the colors of objects, compared to a reference light source with the same color temperature. CRI is an important factor in lighting design, as it affects the perceived color quality and visual clarity of a space.

The standard reference for CRI is CIE 13.3-1995, “Method of Measuring and Specifying Colour Rendering Properties of Light Sources.” This document provides guidelines for measuring and specifying CRI using various instruments and techniques, and it also includes recommendations for minimum CRI values for different lighting applications [3].

CRI is measured on a scale from 0 to 100, with a higher value indicating a better color rendering ability. A CRI of 100 indicates that the light source accurately renders all colors, while a CRI of 0 indicates that all colors appear distorted or washed out under the light.

In practice, CRI is often used as a basis for lighting design and assessment, and various standards and guidelines exist to provide recommendations for CRI values in different applications. For example, the Illuminating Engineering Society (IES) publishes the Lighting Handbook, which includes CRI recommendations for various indoor and outdoor spaces based on factors such as occupancy, task, and visual needs.

However, it is important to note that CRI alone does not fully determine the perceived color quality of a light source. Other factors, such as color temperature and spectral power distribution, can also affect the appearance of colors under the light.

In recent years, the CRI metric has been supplemented with a newer metric called the "General Color Rendering Index" (CRIg), which takes into account the rendering of skin tones, which are important for many applications such as retail, hospitality, and healthcare. The standard reference for CRIg is CIE TN 006:2016, "A Review of the General Color Rendering Index," which provides guidelines for measuring and specifying CRIg using various instruments and techniques.

In summary, CRI is an important measure of a light source's color rendering ability in lighting design and assessment. It is measured on a scale from 0 to 100 and is influenced by factors such as color temperature and spectral power distribution. References such as CIE 13.3-1995 and the IES Lighting Handbook provide guidelines and recommendations for measuring and specifying CRI values in different applications, while the newer metric of CRIg takes into account the rendering of skin tones [3].

1.5 Spectral Power Distribution (SPD)

Spectral power distribution (SPD) is a measure of the intensity of light at each wavelength in the visible spectrum. SPD is an important factor in lighting design, as it affects the perceived color quality and visual clarity of a space, and can also impact visual performance and comfort.

The standard reference for SPD is CIE 15:2004, "Colorimetry." This document provides guidelines for measuring and specifying SPD using various instruments and techniques, and it also includes recommendations for SPD values for different lighting applications.

SPD is typically measured using a spectroradiometer, which can provide a detailed picture of the light source's spectral output. The SPD curve shows the intensity of light at each wavelength in the visible spectrum and can vary widely depending on the light source.

In practice, SPD is often used as a basis for lighting design and assessment, and various standards and guidelines exist to provide recommendations for SPD values in different applications. For example, the Illuminating Engineering Society (IES) publishes the Lighting Handbook, which includes SPD recommendations for various indoor and outdoor spaces based on factors such as occupancy, task, and visual needs.

SPD is also important for understanding the effects of light on human health and well-being. Recent research has shown that the spectral composition of light can affect circadian rhythms, mood, and cognitive performance. For example, "circadian-effective" SPDs have been developed to promote alertness during the day and sleepiness at night, based on the effects of light on the human biological clock.

In summary, SPD is an important measure of the spectral output of a light source in lighting design and assessment. It is typically measured using a spectroradiometer and is influenced by factors such as the light source's spectral power distribution. References such as CIE 15:2004 and the IES Lighting Handbook provide guidelines and recommendations for measuring and specifying SPD values in different applications, while recent research has highlighted the importance of SPD for human health and well-being.

1.6 Flicker

Flicker is a rapid and repeated change in the light output of a lighting system, and it is an important factor in lighting design and assessment. Flicker can cause visual discomfort, affect visual performance, and even trigger seizures in susceptible individuals.

The standard reference for flicker is CIE TN 006:2016, "A Review of Flicker Perception and Flicker Metrics." This document provides guidelines for measuring and specifying flicker using various instruments and techniques, and it also includes recommendations for acceptable flicker levels in different lighting applications.

Flicker is typically measured using a photometer or other specialized instrument, which can provide a detailed analysis of the lighting system's flicker characteristics. Flicker can occur at different frequencies, and the severity of flicker is influenced by factors such as the light source's waveform, dimming level, and power supply.

In practice, flicker is often a concern in applications such as video recording and slow-motion analysis, where even subtle flicker can cause visual artifacts. Flicker is also important for lighting design and assessment, and various standards and guidelines exist to provide recommendations for acceptable flicker levels in different applications. For example, the Illuminating Engineering Society (IES) publishes the Lighting Handbook, which includes flicker recommendations for various indoor and outdoor spaces based on factors such as occupancy, task, and visual needs.

Recent research has also highlighted the potential health effects of flicker. For example, flicker at certain frequencies has been shown to trigger headaches and migraines in some individuals, and it has also been linked to increased symptoms of autism in children.

In summary, flicker is an important factor in lighting design and assessment, and it is typically measured using a photometer or other specialized instrument. References such as CIE TN 006:2016 and the IES Lighting Handbook provide guidelines and recommendations for measuring and specifying flicker levels in different applications, while recent research has highlighted the potential health effects of flicker [4].

1.7 Glare

Glare is a visual sensation caused by excessive and uncontrolled brightness or contrast in a visual field. Glare can cause discomfort, reduce visual performance, and even lead to accidents or injuries.

The standard reference for glare is CIE 117:1995, "Discomfort Glare in Interior Lighting." This document provides guidelines for measuring and specifying glare using various instruments and techniques and it also includes recommendations for acceptable glare levels in different lighting applications [5].

Glare is typically measured using a photometer or other specialized instrument, which can provide a detailed analysis of the lighting system's brightness and contrast levels. Glare can occur in different forms, including disability glare (which reduces visual performance) and discomfort glare (which causes discomfort and annoyance).

In practice, glare is often a concern in applications such as driving, aviation, and outdoor lighting, where excessive brightness or contrast can cause visual impairment or distraction. Glare is also important for lighting design and assessment, and various standards and guidelines exist to provide recommendations for acceptable glare levels in different applications. For example, the Illuminating Engineering Society (IES) publishes the Lighting Handbook, which includes glare recommendations for various indoor and outdoor spaces based on factors such as occupancy, task, and visual needs.

Recent research has also highlighted the potential health effects of glare. For example, glare has been linked to increased risk of falls and accidents in older adults, and it has also been shown to affect sleep quality and circadian rhythms.

In summary, glare is an important factor in lighting design and assessment, and it is typically measured using a photometer or other specialized instrument. References such as CIE 117:1995 and the IES Lighting Handbook provide guidelines and recommendations for measuring and specifying glare levels in different applications, while recent research has highlighted the potential health effects of glare [5].

References

1. Commission Internationale de l'Eclairage. (1987). *CIE 69-1987: Methods of characterizing illuminance meters and luminance meters: Performance, characteristics and specifications*. Commission Internationale de l'Eclairage. https://cie.co.at/publications/methods-characterizing-illuminance-meters-and-luminance-meters-performance
2. Commission Internationale de l'Eclairage. (1986). *CIE 15.2-1986: Colorimetry* (2nd ed.). Commission Internationale de l'Eclairage.
3. Commission Internationale de l'Eclairage. (1995). *CIE 13.3-1995: Method of measuring and specifying colour rendering properties of light sources*. Commission Internationale de l'Eclairage.

4. Commission Internationale de l'Eclairage. (1995). *CIE 117-1995: Discomfort glare in interior lighting*. Commission Internationale de l'Eclairage.
5. IEC 61000-3-3:2013+AMD1:2017+AMD2:2021 CSV. CODE, P., & PRIX, C. Electromagnetic compatibility (EMC)–Part 3-3: Limits–Limitation of voltage changes, voltage fluctuations and flicker in public low-voltage supply systems, for equipment with rated current ≤16 A per phase and not subject to conditional connection Compatibilité électromagnétique (CEM). *Electromagnetic Compatibility (EMC)-Part*, 3-3. https://webstore.iec.ch/publication/68776

Chapter 2
Light and Color Theory

Contents

Light and color theory is the study of how light interacts with objects and how humans perceive color. It is an essential aspect of lighting design, as the color and quality of light can significantly affect the atmosphere and functionality of a space.

There are several key concepts in light and color theory:

- Light Spectrum: The spectrum of light refers to the range of colors that can be seen by the human eye. This spectrum includes all the colors of the rainbow, from red to violet, and also includes ultraviolet and infrared light, which are invisible to the human eye.
- Color Temperature: Color temperature is a measure of the color of light, typically measured in Kelvin (K). It refers to the color of light emitted by a light source, such as a lamp or the sun. Lower color temperatures (below 3000 K) are warm and yellowish, while higher color temperatures (above 5000 K) are cool and bluish [1].
- Color Rendering Index (CRI): CRI is a measure of how well a light source renders colors compared to a reference light source of the same color temperature. The higher the CRI, the more accurate colors will appear under the light source.
- Hue, Saturation, and Brightness: These are the three main properties of color. Hue refers to the actual color, such as red or blue. Saturation refers to the intensity of the color, with highly saturated colors being more vibrant and vivid. Brightness refers to the lightness or darkness of the color [1, 2].

G. Ozenen, *Architectural Interior Lighting*,
https://doi.org/10.1007/978-3-031-49695-0_2

- Contrast: Contrast refers to the difference between the lightest and darkest areas of a space. It is an important aspect of lighting design as it can affect the mood and visual interest of a space.
- Color Psychology: Color psychology is the study of how different colors can affect human emotions and behavior. It is important to consider when designing lighting for different spaces, as the colors used can influence the mood and productivity of the people using the space [1, 3].

2.1 Light Spectrum

Light spectrum refers to the range of wavelengths of electromagnetic radiation that can be detected by the human eye, which is responsible for our perception of color. The visible light spectrum ranges from approximately 400 to 700 nanometers (nm), with violet at the short end and red at the long end. Beyond the visible spectrum are ultraviolet and infrared radiation.

Understanding the light spectrum is important in lighting design because different colors have different effects on mood, perception, and visual acuity. For example, blue light has been shown to improve cognitive performance and alertness, while red light can create a warm and calming ambiance. Additionally, exposure to certain wavelengths of light can affect circadian rhythms and sleep patterns [1].

Light spectrum is also important in the selection of light sources. The spectrum of artificial light sources can vary widely depending on the type of lamp, with some sources emitting light in a narrow range of wavelengths and others emitting a broad spectrum. For example, incandescent bulbs emit a broad spectrum of light, while fluorescent and LED bulbs emit a more limited spectrum [4].

2.2 Color Temperature

Color temperature is an important aspect of light and color theory, particularly in the context of lighting design. It is a measure of the color appearance of light, based on the temperature of an ideal black-body radiator that emits light of a similar hue. Color temperature is measured in Kelvin (K), and typically ranges from warm (lower Kelvin temperatures) to cool (higher Kelvin temperatures) hues.

In lighting design, color temperature is often used to create a specific mood or atmosphere in a space. Warm colors (typically around 2700–3000 K) are often used in residential and hospitality settings to create a cozy and comfortable environment, while cooler colors (around 4000–5000 K) are often used in commercial and institutional settings to create a brighter and more stimulating atmosphere. The choice of color temperature can also affect how colors appear in a space, with cooler colors often making colors appear more vibrant and saturated, while warmer colors can make colors appear more muted and subdued [5].

It is worth noting that color temperature is not the only factor that affects how light appears to the human eye [6]. The color rendering index (CRI) is another important measure, which indicates how well a light source can accurately reproduce colors compared to natural light. A high CRI can be particularly important in settings such as retail and healthcare, where color accuracy is crucial [7].

2.3 Color Rendering Index (CRI)

Color rendering index (CRI) is a metric that measures how accurately a light source can render colors compared to natural sunlight. The CRI score is rated on a scale of 0–100, with a higher score indicating a better color rendering ability of the light source (See Chap. 1).

The CRI is determined by calculating the average of the first eight *R*-values (R1–R8) of the 14 test color samples, with R1 being the most saturated red color and R8 being the most saturated pastel color. The lower the CRI score, the less accurate the colors will appear under that light source. A CRI score of 100 indicates perfect color rendering, where colors appear as they would under natural sunlight [8].

However, there are limitations to the CRI system. For example, it does not accurately assess how colors appear under LED light sources, as LEDs have a very different spectral distribution compared to natural sunlight [1]. Therefore, a new metric, known as the TM-30 method, has been developed to overcome these limitations and provide a more accurate assessment of color rendering under LED lighting [9].

Overall, the CRI is an important consideration when choosing a light source, especially in settings where color accuracy is critical, such as in art galleries or clothing stores.

2.4 Hue, Saturation, and Brightness

Light and color theory includes several fundamental concepts, including hue, saturation, and brightness. These concepts play an essential role in understanding how lighting can be used to create different moods and atmospheres [10].

Hue refers to the actual color of light, such as blue, green, or red. Saturation refers to how pure or intense the color appears, with highly saturated colors appearing more vivid than less saturated ones. Brightness, on the other hand, refers to how much light is emitted, with brighter colors producing more light than darker colors.

These concepts can be visualized using a color wheel, which displays the full range of hues and their relationships to each other. Different hues can create different effects and emotions, with warm hues such as red and orange often associated with energy and excitement, while cool hues such as blue and green are often associated with calm and relaxation [1].

In lighting design, understanding the interplay between hue, saturation, and brightness is crucial for creating the desired atmosphere and achieving the intended visual effects.

2.5 Contrast

Contrast is an important concept in light and color theory, particularly in relation to visual perception and the use of lighting in various applications. Contrast can be defined as the difference between two or more visual elements, such as brightness, color, or texture, which can create visual interest and enhance the clarity and legibility of a scene or object [11].

In lighting design, contrast can be used to create emphasis, highlight specific areas or objects, and direct attention to important details. For example, high contrast lighting can be used in retail spaces to create dramatic visual effects and draw attention to products, while low contrast lighting can be used in galleries or museums to create a more uniform and neutral background that does not distract from the artwork [11, 12].

Contrast can also affect the perceived brightness of a space or object. High contrast lighting can make a space feel brighter and more dynamic, while low contrast lighting can create a more relaxed and subdued atmosphere [12].

In addition to brightness and color contrast, contrast can also be created through texture, shape, and pattern. For example, a textured surface can create visual interest and contrast when illuminated with directional lighting, while a patterned surface can create a sense of movement and depth.

Overall, understanding contrast and its role in visual perception is an important aspect of lighting design and can help create visually appealing and effective lighting schemes.

2.6 Color Psychology

Color psychology refers to the study of how different colors affect human behavior and emotions. It is based on the idea that colors can influence our mood, emotions, and even our physical reactions.

In the context of lighting design, understanding color psychology can help designers choose the right colors for different spaces and applications. For example, warm colors like red, orange, and yellow are associated with energy and warmth, while cool colors like blue and green are associated with calm and relaxation [13].

Here are some common associations with colors and their potential effects on mood:

- Red: Energy, excitement, passion, danger, and urgency
- Orange: Warmth, enthusiasm, creativity, and appetite

- Yellow: Happiness, optimism, attention-grabbing, and caution
- Green: Nature, harmony, balance, and calmness
- Blue: Trust, tranquility, intelligence, and sadness
- Purple: Luxury, creativity, spirituality, and mystery
- Pink: Love, sweetness, femininity, and relaxation
- White: Purity, cleanliness, simplicity, and innocence
- Black: Power, sophistication, elegance, and death

It is important to note that these associations are not universal and can vary based on cultural and personal experiences. However, understanding these basic principles can help lighting designers create spaces that are not only functional but also visually appealing and emotionally engaging [14].

There are some main features that can affect color psychology:

- Hue: The specific color of an object can evoke different emotional responses. For example, blue is often associated with calmness and relaxation, while red is often associated with excitement and energy.
- Saturation: The intensity of a color can also affect its emotional impact. Highly saturated colors can be more energizing and attention-grabbing, while desaturated colors can be more calming and soothing.
- Brightness: The brightness or lightness of a color can also influence its emotional impact. Brighter colors tend to be more energizing and cheerful, while darker colors can be more serious or somber.
- Contrast: The contrast between different colors can also affect their emotional impact. High-contrast combinations can be more stimulating, while low-contrast combinations can be more calming.
- Cultural associations: Different cultures may have different associations with certain colors. For example, in Western cultures, white is often associated with purity and innocence, while in some Eastern cultures, it is associated with mourning and funerals [15].

These features are not exhaustive, but they are some of the key factors that can affect the psychological impact of color.

References

1. Wotton, E. (2000). The IESNA lighting handbook and office lighting. *Lighting, 14,* pp.10–11
2. Sumartojo, S. (Ed.). (2022). *Lighting design in shared public spaces*. Routledge.
3. Cuttle, C. (2018). *Lighting design: A perception-based approach* (2nd ed.). Routledge.
4. National Institute of Standards and Technology. (2020). *Electromagnetic spectrum*. Retrieved from https://www.nist.gov/image/06phy009emspec2hrjpg
5. Lighting Research Center. *Understanding white light source color rendering and appearance*. Rensselaer Polytechnic Institute. https://www.lrc.rpi.edu/programs/solidstate/colorresearch.asp
6. U.S. Department of Energy. *Color and spectrum*. https://www.energy.gov/eere/ssl/color-and-spectrum

7. Lighting Research Center. *Developing a predictive life test for LED systems*. Rensselaer Polytechnic Institute. https://www.lrc.rpi.edu/programs/solidstate/LEDSystemLife.asp
8. ANSI/IES RP-16-17, Nomenclature and Definitions for Illuminating Engineering.
9. Karlen, M., Spangler, C., & Benya, J. R. (2017). *Lighting design basics*. Wiley.
10. Bellia, L., Bisegna, F., & Spada, G. (2011). Lighting in indoor environments: Visual and non-visual effects of light sources with different spectral power distributions. *Building and Environment, 46*(10), 1984–1992.
11. Boyce, P. R. (2014). *Human factors in lighting*. CRC Press.
12. Livingston, J. (2021). *Designing with light: The art, science, and practice of architectural lighting design*. Wiley.
13. Mikellides, B. (2012). Colour psychology: The emotional effects of colour perception. In *Colour design* (pp. 105–128). Woodhead Publishing.
14. Elliot, A. J., Maier, M. A., Moller, A. C., Friedman, R., & Meinhardt, J. (2007). Color and psychological functioning: The effect of red on performance attainment. *Journal of Experimental Psychology: General, 136*(1), 154.
15. Stone, N. J. (2008). Human factors and education: Evolution and contributions. *Human Factors, 50*(3), 534–539.

Chapter 3
Lighting Fundamentals and Design Principles

Contents

Lighting fundamentals and design principles are important in creating effective lighting solutions that meet the needs of users while minimizing energy consumption and environmental impact. Here are some of the key fundamentals and principles to keep in mind:

- The Lighting Triangle: The lighting triangle is a basic concept in lighting design that involves balancing the three primary elements of lighting – ambient lighting, accent lighting, and task lighting – to create a visually comfortable and balanced environment. Ambient lighting provides general illumination, while accent lighting highlights specific objects or areas. Task lighting is used for specific tasks such as reading or cooking [1].
- Lighting Layers: Lighting layers are used to create depth and interest in a space by combining different types of lighting, including ambient, task, and accent lighting. By using layers, lighting designers can create visual interest while providing the appropriate levels of illumination for specific tasks.
- Lighting Quality: Lighting quality refers to the visual experience of a space, including the color, brightness, and direction of light. A well-designed lighting

G. Ozenen, *Architectural Interior Lighting*,
https://doi.org/10.1007/978-3-031-49695-0_3

system should create an environment that is comfortable, visually appealing, and supports the needs of the users [1, 2].

- Lighting Intensity: Lighting intensity refers to the brightness of the light and is measured in lumens. A lighting design should provide sufficient intensity to meet the needs of the users, while avoiding over-illumination and wasting energy.
- Color Rendering: Color rendering refers to the ability of a lighting system to accurately reproduce colors. A high-quality lighting system should provide accurate color rendering to support visual tasks and create a visually pleasing environment.
- Glare Control: Glare occurs when there is excessive contrast between the light source and the surrounding environment. A well-designed lighting system should minimize glare to create a comfortable and visually pleasing environment [3].
- Energy Efficiency: Energy efficiency is an important consideration in lighting design, as it helps to minimize the environmental impact of lighting systems. By using energy-efficient lighting technologies and strategies such as daylight harvesting and lighting controls, lighting designers can reduce energy consumption and minimize the environmental impact of lighting systems.

3.1 The Lighting Triangle

The lighting triangle is a fundamental concept in lighting design that involves the interplay of three key elements: light source, light fixture, and lighting design. It is a useful tool for lighting designers to create a balanced and effective lighting scheme that meets the functional and aesthetic requirements of a space.

The light source is the origin of light, and it can be natural or artificial. Natural light sources include the sun and sky, while artificial light sources include electric lamps and luminaires. Light fixtures are the devices that hold and distribute the light from the source. These fixtures can be wall-mounted, ceiling-mounted, or freestanding, and can be designed for specific purposes, such as task lighting or ambient lighting. Finally, the lighting design is the arrangement of light sources and fixtures to create a desired lighting effect [4].

To achieve a successful lighting design, the lighting triangle must be balanced. This means that the amount of light produced by the light source must match the needs of the space, the light fixtures must be appropriate for the light source and the space, and the lighting design must be effective in achieving the desired lighting effect while considering factors such as energy efficiency and maintenance requirements [5, 6].

Other factors that can affect the lighting triangle include color temperature, color rendering index (CRI), and light distribution. Color temperature refers to the perceived warmth or coolness of a light source and is measured in Kelvin (K). CRI measures how accurately a light source renders colors compared to natural light and is rated on a scale from 0 to 100. Light distribution refers to how evenly light is distributed throughout a space and can be affected by the type of light fixture used and its placement [6].

3.2 Lighting Layers

Lighting layers is a fundamental concept in lighting design that refers to the different types of lighting needed to create a visually comfortable and stimulating environment. The concept is based on the idea that a space needs different layers of lighting to be properly illuminated, and each layer serves a specific purpose. There are three main layers of lighting: ambient lighting, task lighting, and accent lighting.

- Ambient Lighting: Ambient lighting is the general illumination of a space. This layer of lighting provides a uniform level of illumination throughout the space and serves as the base for the other layers of lighting. It can be achieved through the use of ceiling-mounted fixtures, recessed lighting, or even natural light from windows.
- Task Lighting: Task lighting is focused lighting that is used to illuminate specific areas where activities such as reading, cooking, or working take place. This layer of lighting provides brighter and more localized light to help users perform specific tasks with ease. Task lighting can be achieved through the use of table lamps, floor lamps, or under-cabinet lighting [7].
- Accent Lighting: Accent lighting is used to highlight specific features of a space, such as artwork, architectural details, or decorative objects. This layer of lighting adds visual interest and drama to a space and can be achieved through the use of track lighting, wall-mounted fixtures, or directional spotlights [8].

When these layers of lighting are combined, they create a balance of illumination that enhances the functionality and aesthetic appeal of a space. Effective lighting design requires an understanding of the space, the activities that take place within it, and the desired mood or ambiance.

3.2.1 Ambient Lighting

Ambient lighting is the base layer of lighting in a space, providing overall illumination to create a comfortable and inviting atmosphere. This type of lighting is also known as general lighting, and it is often achieved through the use of overhead fixtures, such as recessed lights, ceiling-mounted fixtures, or pendant lights [7, 8].

In addition to providing general illumination, ambient lighting can also enhance the visual appeal of a space. It can be used to highlight architectural features, provide a sense of depth, or create a particular mood. For example, warm ambient lighting can create a cozy and inviting atmosphere in a living room, while cool ambient lighting can create a refreshing and energizing atmosphere in a bathroom or kitchen [7, 8].

When designing ambient lighting, it is important to consider the size and shape of the space, as well as the needs of the people who will be using it. Some common factors to consider include the color and reflectivity of the walls, the height of the ceiling, and the amount of natural light entering the space [9].

To ensure that ambient lighting is energy-efficient, it is important to choose fixtures that are appropriate for the size of the space and the desired level of illumination. LED bulbs are a popular choice for ambient lighting, as they are long-lasting and energy-efficient [9].

3.2.2 *Task Lighting*

Task lighting is a specific type of lighting used to provide illumination for specific tasks or activities, such as reading, writing, or cooking. It is an essential component of lighting design as it helps to increase productivity and reduce eye strain by providing sufficient and adjustable light levels in a particular workspace [10].

Task lighting can be achieved using various types of fixtures, including desk lamps, floor lamps, under-cabinet lights, and pendant lights. These fixtures should be placed in a way that illuminates the task area without causing glare or shadows. For example, a desk lamp should be positioned on the opposite side of the dominant hand to prevent casting a shadow on the work surface [10, 11].

The design of task lighting should also take into consideration the color rendering index (CRI) of the light source. CRI is a measure of a light source's ability to accurately render colors, and it is crucial for tasks that require color discrimination, such as painting or sewing. A higher CRI value indicates better color rendering, and a CRI of at least 80 is recommended for most tasks [10, 12].

Furthermore, task lighting can be integrated into a lighting control system to provide additional energy savings. For example, occupancy sensors can be used to turn off task lighting when the area is unoccupied, and dimming controls can be used to adjust light levels according to the task and time of day.

3.2.3 *Accent Lighting*

Accent lighting is a type of lighting that is used to draw attention to a particular object or area in a space. The purpose of accent lighting is to create visual interest and enhance the overall aesthetic of a space. This type of lighting is often used in museums, galleries, retail stores, and homes to highlight art, displays, or architectural features [13].

There are several types of accent lighting, including the following:

- Spotlights: These are directional lights that are used to illuminate a specific object or area. Spotlights can be adjusted to change the size and direction of the beam of light.
- Wall Washers: These lights are mounted on a wall and are used to wash a wall or large area with light. They create a soft and even illumination that is ideal for highlighting artwork or architectural details.
- Picture Lights: These lights are typically mounted above a piece of artwork or photograph and are used to provide a focused beam of light on the subject.
- Under-Cabinet Lighting: This lighting is often used in kitchens to provide additional light for food preparation and cooking. It can also be used to highlight decorative items on shelves or cabinets [13, 14].

When designing accent lighting, it is important to consider the color temperature and color rendering of the light source to ensure that the illuminated object or area is presented in the best possible way. It is also important to use the appropriate intensity of light to create the desired effect without overpowering the space.

3.3 Lighting Quality

Lighting quality is a crucial aspect of lighting design that directly affects visual comfort, mood, and health. Good lighting quality involves creating a lighting environment that provides adequate illumination, visual interest, and uniformity while minimizing glare and discomfort. It is important to consider both the technical aspects of lighting, such as color temperature and CRI, as well as the subjective experience of the space and its occupants.

Here are some key factors that contribute to lighting quality in lighting design:

- Illumination: Illumination refers to the amount of light falling on a surface and is usually measured in lux or foot-candles. Adequate illumination levels should be provided for different activities and areas based on their requirements, such as reading, task performance, or ambiance [15].
- Color temperature and CRI: Color temperature and CRI are important technical factors that affect the color appearance of objects in a space. Color temperature is measured in Kelvin (K) and describes the warmth or coolness of the light source. CRI, on the other hand, is a measure of how accurately a light source can render colors compared to natural daylight. Higher CRI values result in more accurate color representation [16].
- Uniformity: Uniformity refers to the consistency of illumination levels across a space. A good lighting design should aim to minimize variations in light levels that could cause visual discomfort or distraction [17].
- Glare: Glare occurs when excessively bright light sources or reflections cause visual discomfort or reduce visibility. Glare can be reduced by using proper shielding or diffusing techniques.

- Shadow: Shadow plays an important role in lighting design by creating depth and visual interest in a space. Properly placed lighting can enhance the natural contours of a space, making it more visually appealing and engaging.
- Flicker: Flicker refers to the rapid and repeated changes in light intensity that can cause visual discomfort or even health issues. Good lighting design should minimize flicker by selecting high-quality lighting products and avoiding cheap or poorly designed fixtures [18, 19].
- Circadian rhythm: Circadian rhythm refers to the natural biological rhythms that regulate our sleep-wake cycle and other bodily functions. Proper lighting design can support these rhythms by providing lighting that mimics natural daylight and adjusts throughout the day [20, 21].

3.4 Lighting Intensity

Lighting intensity is an important aspect of lighting design that refers to the amount of light output from a lighting system. It is typically measured in lux (lx) or foot-candles (fc) and is influenced by several factors such as the distance between the light source and the object, the reflectance of the surfaces, and the type of light source used [22].

The lighting intensity required for a particular application depends on various factors, including the visual task, the age and visual acuity of the occupants, and the type of space being illuminated. The Illuminating Engineering Society (IES) provides recommended lighting levels for different applications in its Lighting Handbook, which is widely used as a reference in the lighting industry.

Proper lighting intensity can improve visual acuity, reduce eye strain, and enhance the overall visual experience. However, excessive lighting intensity can cause discomfort and glare, which can negatively affect visual performance and cause discomfort for occupants [23].

To ensure appropriate lighting intensity, lighting designers use various techniques, including selecting appropriate light sources, calculating light levels based on the visual task and space being illuminated, and designing the lighting layout to minimize glare and provide even illumination.

3.5 Color Rendering

Color rendering is a critical aspect of lighting design, as it affects how colors appear under artificial lighting. Color rendering refers to the ability of a light source to reveal the true colors of objects as they would appear under natural daylight. In other words, it is a measure of how accurately a light source can reproduce colors.

The Color Rendering Index (CRI) is a metric used to quantify color rendering. The CRI scale ranges from 0 to 100, with 100 being the highest possible score. A CRI score of 80 or above is typically recommended for most applications, although some applications, such as art galleries or fashion showrooms, may require a CRI score of 90 or higher [24].

Several factors can affect color rendering, including the light source's spectral power distribution, the illuminance level, and the color temperature. Light sources with a higher CRI typically have a smoother spectral power distribution, which allows them to render colors more accurately. Higher illuminance levels also improve color rendering, as they provide more light to reveal colors. Finally, the color temperature of the light source can affect color rendering, as some colors may appear different under different color temperatures [25].

In addition to color rendering, lighting designers must also consider other aspects of lighting quality, such as glare control, light distribution, and visual comfort. Proper lighting design can enhance the visual environment, promote safety, and improve the overall experience for occupants.

3.6 Glare Control

Glare is defined as excessive brightness or contrast in the visual field that causes discomfort or reduces the ability to see. Glare in lighting design can be caused by direct or reflected light sources, and it can reduce visual comfort, visibility, and performance. Therefore, glare control is an essential aspect of lighting design that aims to minimize the negative effects of glare on the occupants. Here are some key principles and strategies for glare control in lighting design:

- Limit the luminance of light sources: The luminance of a light source is the amount of light emitted per unit solid angle and per unit projected area. High-luminance light sources can cause discomfort and disability glare. Therefore, lighting designers should select light sources with appropriate luminance levels and use diffusing or shielding devices to reduce the luminance of the light sources [26].
- Control the direction of light: Direct light sources that are visible from the viewing area can cause glare. Therefore, lighting designers should control the direction of light and use indirect or semi-indirect lighting fixtures that reflect light off ceilings, walls, or other surfaces to reduce direct glare.
- Use appropriate reflectance values for surfaces: The reflectance value of a surface is the amount of light reflected from the surface compared to the amount of light received by the surface. Low-reflectance surfaces can reduce visual comfort and increase the amount of reflected glare. Therefore, lighting designers should use appropriate reflectance values for surfaces and avoid high-gloss surfaces that can cause specular reflections [26, 27].

- Use appropriate lighting distribution: The lighting distribution refers to the pattern of light on the task and surrounding surfaces. Lighting designers should use appropriate lighting distribution that provides uniform and comfortable illumination on the task and reduces glare on the surrounding surfaces.
- Use appropriate lighting control systems: Lighting control systems such as dimmers, timers, and sensors can help reduce glare by providing the right amount of light at the right time. For example, dimmers can adjust the light level to reduce glare, while occupancy sensors can turn off the lights when the space is unoccupied to reduce unnecessary lighting [28].

Glare control is an essential aspect of lighting design that aims to provide comfortable and efficient illumination while minimizing the negative effects of glare on the occupants. By considering the above principles and strategies, lighting designers can achieve optimal glare control in their designs.

3.7 Energy Efficiency

Energy efficiency is a critical consideration in lighting design, as it can significantly impact the operational cost and environmental impact of a lighting system. The adoption of energy-efficient lighting technologies can lead to reduced energy consumption, lower greenhouse gas emissions, and lower electricity bills. Energy efficiency is a fundamental aspect of lighting quality, and it is essential to consider energy performance metrics, such as efficacy, luminous flux, and power consumption, when designing lighting systems [29].

Several energy-efficient lighting technologies have emerged in recent years, including LED, OLED, and CFL lamps, which have significantly higher efficacy and longer lifespan than traditional incandescent lamps. LED technology, in particular, has become the dominant lighting technology due to its superior energy efficiency, versatility, and durability. LED lamps are typically three to four times more efficient than incandescent lamps, have a lifespan of up to 50,000 h, and are available in various color temperatures and color rendering indices [29, 30].

In addition to using energy-efficient lamps, designers can also optimize energy efficiency by selecting appropriate lighting controls and strategies. For example, dimming, occupancy sensing, daylight harvesting, and scheduling controls can all contribute to reducing energy consumption by ensuring that lighting is only used when and where it is needed. Furthermore, networked lighting controls can enable more granular control of lighting systems, allowing designers to optimize lighting levels for specific tasks, areas, and times of the day [31].

Energy efficiency is a critical consideration in lighting design, and designers must consider the energy performance of lighting systems to optimize their environmental and economic impact.

References

1. Eley, C., Tolen, T., & Benya, J. R. (1992). *Lighting fundamentals handbook: Lighting fundamentals and principles for utility personnel* (No. EPRI-TR-101710). Electric Power Research Institute; Eley (Charles) Associates; Luminae Souter Lighting Design.
2. Dunham, R. E. (2015). *Stage lighting: Fundamentals and applications*. CRC Press.
3. Van Bommel, W. (2014). *Road lighting: Fundamentals, technology and application*. Springer.
4. Hemming, S. (2009). Use of natural and artificial light in horticulture-interaction of plant and technology. In *Proceedings of the VIth international symposium on light in horticulture* (pp. 25–35). ISHS.
5. Gaston, K. J., & Sánchez de Miguel, A. (2022). Environmental impacts of artificial light at night. *Annual Review of Environment and Resources, 47*, 373–398.
6. Gaston, K. J., Visser, M. E., & Hölker, F. (2015). The biological impacts of artificial light at night: The research challenge. *Philosophical Transactions of the Royal Society B: Biological Sciences, 370*(1667), 20140133.
7. Bloom, R., Zamalloa, M. Z., & Pai, C. (2019). LuxLink: Creating a wireless link from ambient light. In *Proceedings of the 17th conference on embedded networked sensor systems* (pp. 166–178). ACM.
8. Wang, T. T., Guo, K., Hu, X. M., Liang, J., Li, X. D., Zhang, Z. F., & Xie, J. (2020). Label-free colorimetric detection of urine glucose based on color fading using smartphone ambient-light sensor. *Chemosensors, 8*(1), 10.
9. Canazei, M., Dehoff, P., Staggl, S., & Pohl, W. (2014). Effects of dynamic ambient lighting on female permanent morning shift workers. *Lighting Research & Technology, 46*(2), 140–156.
10. Juslén, H., Wouters, M., & Tenner, A. (2007). The influence of controllable task-lighting on productivity: A field study in a factory. *Applied Ergonomics, 38*(1), 39–44.
11. Davis, R. G., & Garza, A. (2002). Task lighting for the elderly. *Journal of the Illuminating Engineering Society, 31*(1), 20–32.
12. Newsham, G., Arsenault, C., Veitch, J., Tosco, A. M., & Duval, C. (2005). Task lighting effects on office worker satisfaction and performance, and energy efficiency. *LEUKOS, 1*(4), 7–26.
13. Gunter, H. (2004). Accent lighting helps designers change a room's atmosphere. *Hotel & Motel Management, 219*(14), 143–143.
14. Van Kemenade, I. J., & Reker, J. (1988). Beam characteristics for accent lighting. *Journal of the Illuminating Engineering Society, 17*(2), 118–130.
15. Ritschel, T., Dachsbacher, C., Grosch, T., & Kautz, J. (2012). The state of the art in interactive global illumination. In *Computer graphics forum* (Vol. 31, No. 1, pp. 160–188). Blackwell Publishing Ltd.
16. Lin, D., Zhong, P., & He, G. (2017). Color temperature tunable white LED cluster with color rendering index above 98. *IEEE Photonics Technology Letters, 29*(12), 1050–1053.
17. Runkle, E. R. I. K. (2017). The importance of light uniformity. *GPNMAG.COM*, 2.
18. Ticleanu, C., & Littlefair, P. (2019). *Quality indoor lighting for comfort, health, well-being and productivity*. BRE Trust.
19. Dodo, Y. A., Kandar, M. Z., Girei, M. B., & Abdul, A. Y. (2012). Comparative analysis of effect of psychological factors on visual comfort in a green and conventional office building. In *2012 2nd international conference on biotechnology and environment management, IPCBEE* (Vol. 42). IACSIT Press.
20. Tähkämö, L., Partonen, T., & Pesonen, A. K. (2019). Systematic review of light exposure impact on human circadian rhythm. *Chronobiology International, 36*(2), 151–170.
21. LeGates, T. A., Fernandez, D. C., & Hattar, S. (2014). Light as a central modulator of circadian rhythms, sleep and affect. *Nature Reviews Neuroscience, 15*(7), 443–454.
22. Shafiq, I., Hussain, S., Raza, M. A., Iqbal, N., Asghar, M. A., Ali, R. A. Z. A., et al. (2021). Crop photosynthetic response to light quality and light intensity. *Journal of Integrative Agriculture, 20*(1), 4–23.

23. Deep, A., Schwean-Lardner, K., Crowe, T. G., Fancher, B. I., & Classen, H. L. (2010). Effect of light intensity on broiler production, processing characteristics, and welfare. *Poultry Science, 89*(11), 2326–2333.
24. Zhang, J., Xu, G., Tao, F., Zeng, G., Zhang, M., Yang, Y., et al. (2019). Highly efficient semitransparent organic solar cells with color rendering index approaching 100. *Advanced Materials, 31*(10), 1807159.
25. Yang, X., Sui, L., Wang, B., Zhang, Y., Tang, Z., Yang, B., & Lu, S. (2021). Red-emitting, self-oxidizing carbon dots for the preparation of white LEDs with super-high color rendering index. *Science China Chemistry, 64*(9), 1547–1553.
26. Sun, Y., Liu, X., Qu, W., Cao, G., & Zou, N. (2020). Analysis of daylight glare and optimal lighting design for comfortable office lighting. *Optik, 206*, 164291.
27. Hopkinson, R. G. (1940). Discomfort glare in lighted streets. *Transactions of the Illuminating Engineering Society, 5*(1–9_IEStrans), 1–32.
28. Bellia, L., Cesarano, A., Iuliano, G. F., & Spada, G. (2008). Daylight glare: A review of discomfort indexes. In *Visual quality and energy efficiency in indoor lighting: Today for tomorrow*. IRIS Unina.
29. Loe, D. L. (2009). Energy efficiency in lighting—Considerations and possibilities. *Lighting Research & Technology, 41*(3), 209–218.
30. Sozer, H. (2010). Improving energy efficiency through the design of the building envelope. *Building and Environment, 45*(12), 2581–2593.
31. Dai, Q., Huang, Y., Hao, L., Lin, Y., & Chen, K. (2018). Spatial and spectral illumination design for energy-efficient circadian lighting. *Building and Environment, 146*, 216–225.

Chapter 4
Basic Principles of Lighting

Contents

4.1 Lighting Design

Lighting design is the process of creating lighting systems that meet the functional and aesthetic needs of a space or application, while also considering factors such as energy efficiency, sustainability, and human factors. Effective lighting design involves a balance of technical and artistic considerations, as well as an understanding of the physiological and psychological effects of light on human beings [1, 2].

G. Ozenen, *Architectural Interior Lighting*,
https://doi.org/10.1007/978-3-031-49695-0_4

The primary objective of lighting design is to provide sufficient and appropriate illumination to enable visual tasks to be performed safely, comfortably, and efficiently, while also enhancing the visual appearance of the environment [3]. Achieving this objective requires careful consideration of several key factors, including illuminance, luminance, color temperature, color rendering index (CRI), spectral power distribution (SPD).

Moreover, there are some other factors which are described in the following sections.

4.1.1 *Energy Efficiency*

Lighting design should be energy efficient, using the most efficient light sources and control systems appropriate for the space.

Energy efficiency is an important aspect of lighting design, as lighting systems can account for a significant portion of a building's energy consumption. There are various ways to improve the energy efficiency of lighting systems, including the use of more efficient light sources, such as LED, and the implementation of lighting controls to reduce energy consumption when lighting is not needed [2, 3].

One approach to designing energy-efficient lighting systems is to use a performance-based approach, which focuses on designing lighting systems that meet specific performance criteria, rather than simply meeting minimum energy code requirements. This approach allows for more flexibility and creativity in the design process, while still ensuring that the lighting system is energy-efficient.

Another approach to improving energy efficiency in lighting design is through the use of daylighting systems, which harness natural daylight to provide lighting for interior spaces, reducing the need for artificial lighting. Daylighting systems can include skylights, windows, and light shelves, among other components, and can be designed to provide optimal levels of natural light while minimizing glare and heat gain.

The use of lighting controls, such as occupancy sensors and daylight sensors, can also improve energy efficiency in lighting design by reducing energy consumption when lighting is not needed. For example, occupancy sensors can automatically turn off lights in unoccupied rooms, while daylight sensors can dim or turn off artificial lighting in response to available daylight [4].

4.1.2 *Sustainability*

Lighting design should be sustainable, considering factors such as the use of renewable energy sources, the reduction of light pollution, and the minimization of environmental impact.

Sustainability in lighting design refers to the practice of designing lighting systems that are environmentally and socially responsible, while also meeting the functional and aesthetic requirements of the space. This involves considering the entire lifecycle of the lighting system, from the production and disposal of materials to energy consumption during use.

One approach to sustainable lighting design is to use energy-efficient light sources, such as LEDs or compact fluorescent lamps, and to optimize the lighting design to reduce energy consumption while still meeting the lighting needs of the space. Another approach is to incorporate natural daylight into the lighting design, which can reduce the need for electric lighting and provide a connection to the outdoors.

Sustainable lighting design also considers the materials and manufacturing processes used in producing lighting fixtures, as well as the environmental impact of their disposal. Designers may choose to use materials with a lower environmental impact, such as recycled or renewable materials, and to design fixtures that are easily disassembled and recyclable.

In addition to environmental sustainability, lighting design can also contribute to social sustainability by creating spaces that are comfortable, safe, and visually appealing. For example, lighting can be used to enhance wayfinding and create a sense of security in outdoor spaces [5].

4.1.3 Human Factors

Lighting design should consider the physiological and psychological effects of light on human beings, such as circadian rhythm, alertness, and mood.

Aesthetics: Lighting design should enhance the visual appearance of the space, using appropriate color, brightness, and contrast.

Human factors are an important consideration in lighting design as lighting affects human health, well-being, and productivity. Lighting can affect human circadian rhythms, mood, and visual comfort, and it is important to consider these factors in lighting design. Here are some details about human factors in lighting design with academic references:

- Circadian rhythms: Light plays an important role in regulating human circadian rhythms, which affect sleep-wake cycles, alertness, and mood. Research has shown that exposure to light with high color temperature and high illuminance levels in the morning can help regulate circadian rhythms and improve sleep quality.
- Visual comfort: Lighting design should also take into account visual comfort, which includes factors such as glare, contrast, and color rendering. High levels of glare and poor color rendering can cause discomfort, eye strain, and headaches and can also affect task performance.
- Mood and well-being: Light can also affect mood and well-being, with studies showing that exposure to bright light can improve mood and reduce symptoms of

depression and anxiety. Lighting design can incorporate features such as dynamic lighting and biophilic design to enhance these effects.

- Age and vision: The aging population has different lighting needs than younger people, and lighting design should take into account factors such as reduced visual acuity, contrast sensitivity, and color perception. Research has shown that lighting interventions can improve visual performance and reduce the risk of falls in older adults [6].

Effective lighting design requires a multi-disciplinary approach, involving collaboration between lighting designers, architects, engineers, and other professionals. There are several guidelines and standards available to help ensure that lighting design is effective, safe, and sustainable [6, 7].

4.2 Lighting Controls

Devices and systems used to adjust the lighting levels or color of a lighting system, typically include dimmers, switches, timers, and sensors.

Lighting controls refer to devices and systems that allow users to adjust and manage the lighting levels and conditions in a space. The main purpose of lighting controls is to reduce energy consumption and operating costs while enhancing the visual and non-visual benefits of lighting. Effective lighting controls can also improve the occupant experience, safety, and productivity, and support the health and well-being of the occupants.

The key components of lighting controls include sensors, switches, dimmers, timers, and control systems. These components can be used to control the amount, direction, color, and quality of light in a space, based on the occupancy, daylight, and task requirements.

There are several types of lighting controls available, including the following.

4.2.1 Occupancy Sensors

These sensors detect the presence or absence of people in a space and automatically turn the lights on or off, or adjust the lighting levels accordingly. They are commonly used in areas such as restrooms, hallways, and storage rooms.

4.2.2 Daylight Sensors

These sensors measure the amount of natural light entering a space and adjust the electric lighting levels to maintain a desired light level. They are commonly used in spaces such as offices, classrooms, and retail stores.

4.2.3 *Dimmers*

These devices allow users to adjust the brightness of a light source, providing flexibility and energy savings. They are commonly used in spaces such as restaurants, conference rooms, and homes.

4.2.4 *Timeclocks*

These devices allow users to program lighting schedules based on occupancy patterns, time of day, and season. They are commonly used in spaces such as parking lots, outdoor areas, and sports facilities.

4.2.5 *Networked Control Systems*

These systems allow users to remotely monitor and manage lighting levels and energy consumption across multiple spaces and buildings. They are commonly used in large commercial, institutional, and industrial settings.

Effective lighting control design requires a thorough understanding of the space's lighting needs and occupant behavior, as well as an understanding of the available technologies and energy codes and standards. Lighting control design should also consider the integration with other building systems such as HVAC, security, and audiovisual [8–10].

4.3 Lighting Standards and Regulations

Various codes, standards, and regulations exist to ensure that lighting systems are safe, effective, and sustainable and to promote energy efficiency, human health, and environmental protection.

Lighting standards and regulations are guidelines and requirements that govern the design, installation, and operation of lighting systems in buildings, outdoor spaces, and transportation facilities. The main objectives of these standards and regulations are to ensure the safety, health, energy efficiency, and environmental sustainability of lighting systems, as well as to support the visual and non-visual needs of the occupants and users.

The most widely recognized organizations that establish and maintain lighting standards and regulations are the Illuminating Engineering Society (IES), the International Commission on Illumination (CIE), and the International Electrotechnical Commission (IEC). These organizations develop and publish a wide range of technical reports, guidelines, codes, and standards that cover various aspects of lighting design, technology, measurement, and performance.

Following are some of the most important lighting standards and regulations.

4.3.1 Energy Codes

These codes establish minimum energy efficiency requirements for lighting systems, in order to reduce energy consumption, greenhouse gas emissions, and operating costs. Examples of energy codes include ASHRAE 90.1, IECC, and Title 24.

Energy codes are regulations and standards that are aimed at reducing the energy consumption and carbon emissions of buildings. Here are some details about energy codes, along with academic references:

- Purpose: The purpose of energy codes is to set minimum requirements for the energy performance of buildings, including their heating, cooling, and lighting systems. The goal is to reduce energy consumption, decrease greenhouse gas emissions, and improve the overall energy efficiency of the building sector.
- Types of codes: There are many different energy codes used around the world, including the International Energy Conservation Code (IECC) in the United States, the National Building Code of Canada, and the European Union's Energy Performance of Buildings Directive. These codes are regularly updated to reflect advances in building technology and changes in energy policy.
- Compliance: Compliance with energy codes is typically enforced through inspections and permits. Builders and property owners are required to demonstrate that their buildings meet the energy code requirements in order to obtain permits and occupancy certificates.
- Impacts: Energy codes can have significant impacts on building energy use and greenhouse gas emissions. Studies have shown that buildings constructed to energy code standards consume less energy than non-code-compliant buildings and that energy code compliance can result in significant cost savings for building owners and occupants.
- Challenges: Implementing energy codes can also present challenges, such as lack of awareness and knowledge among building professionals, resistance from the construction industry, and the need for enforcement and monitoring [11–13].

4.3.2 Safety Codes

These codes establish minimum safety requirements for lighting systems, in order to prevent electrical shocks, fires, and other hazards. Examples of safety codes include NFPA 70, IEC 60364, and BS 7671.

Safety codes are regulations and standards that are aimed at ensuring the safety of people and property in buildings. Here are some details about safety codes, along with academic references:

- Purpose: The purpose of safety codes is to set minimum requirements for the safety and health of people who use buildings, as well as for the protection of property. Safety codes cover a wide range of topics, including fire safety, electrical safety, structural safety, and accessibility.
- Types of codes: There are many different safety codes used around the world, including the National Fire Protection Association (NFPA) codes in the United States, the Canadian Electrical Code, and the International Building Code (IBC). These codes are regularly updated to reflect advances in building technology and changes in safety policy.
- Compliance: Compliance with safety codes is typically enforced through inspections and permits. Builders and property owners are required to demonstrate that their buildings meet the safety code requirements in order to obtain permits and occupancy certificates.
- Impacts: Safety codes can have significant impacts on the safety and health of building occupants and the protection of property. Studies have shown that buildings constructed to safety code standards are less likely to experience fires and other safety hazards and that compliance with safety codes can result in significant cost savings for building owners and occupants.
- Challenges: Implementing safety codes can also present challenges, such as lack of awareness and knowledge among building professionals, resistance from the construction industry, and the need for enforcement and monitoring [14–18].

4.3.3 Accessibility Codes

These codes establish minimum accessibility requirements for lighting systems, in order to ensure that people with disabilities can use and benefit from the lighting. Examples of accessibility codes include ADA, EN 12464, and AS 1428.

Accessibility codes are regulations and standards that are aimed at ensuring that buildings and facilities are accessible to people with disabilities. Here are some details about accessibility codes, along with academic references:

- Purpose: The purpose of accessibility codes is to ensure that people with disabilities have equal access to buildings and facilities and that they are not discriminated against on the basis of their disability. Accessibility codes cover a wide range of topics, including physical accessibility, communication accessibility, and accessibility of digital and electronic technologies.
- Types of codes: There are many different accessibility codes used around the world, including the Americans with Disabilities Act (ADA) in the United States, the Accessibility for Ontarians with Disabilities Act (AODA) in Canada, and the

Disability Discrimination Act (DDA) in the United Kingdom. These codes are regularly updated to reflect advances in accessibility technology and changes in accessibility policy.

- Compliance: Compliance with accessibility codes is typically enforced through inspections and penalties for non-compliance. Builders and property owners are required to demonstrate that their buildings meet the accessibility code requirements in order to obtain permits and occupancy certificates.
- Impacts: Accessibility codes can have significant impacts on the ability of people with disabilities to participate fully in society. Studies have shown that buildings constructed to accessibility code standards are more accessible and usable for people with disabilities and that compliance with accessibility codes can result in significant cost savings for building owners and occupants.
- Challenges: Implementing accessibility codes can also present challenges, such as lack of awareness and knowledge among building professionals, resistance from the construction industry, and the need for enforcement and monitoring [19–22].

4.3.4 Environmental Codes

These codes establish minimum environmental requirements for lighting systems, in order to reduce the impact of lighting on natural ecosystems, such as wildlife, plants, and skies. Examples of environmental codes include IDA, LEED, and BREEAM.

Environmental codes refer to regulations and guidelines that address the impact of buildings and their lighting systems on the environment. These codes aim to reduce energy consumption, carbon emissions, and waste production, among other environmental concerns. Some examples of environmental codes related to lighting systems include the following:

- Leadership in Energy and Environmental Design (LEED): LEED is a rating system developed by the US Green Building Council to evaluate the sustainability of buildings. It includes various criteria related to lighting, such as energy efficiency, lighting controls, and daylighting.
- Energy Star: Energy Star is a voluntary program by the US Environmental Protection Agency that promotes energy-efficient products, including lighting fixtures and bulbs. Energy Star-certified products meet certain energy efficiency criteria and can help reduce energy consumption and greenhouse gas emissions.
- International Energy Conservation Code (IECC): The IECC is a model energy code developed by the International Code Council to promote energy-efficient buildings. The code includes requirements related to lighting systems, such as lighting power density limits, occupancy sensors, and daylighting controls.

- California Title 24: Title 24 is a set of energy standards developed by the California Energy Commission for new construction and major renovations. The standards include requirements related to lighting, such as lighting power density limits, automatic controls, and daylighting [23–26].

4.3.5 Performance Standards

These standards establish minimum performance requirements for lighting systems, in order to ensure that they meet certain criteria of quality, efficacy, and functionality. Examples of performance standards include ENERGY STAR, DLC, and Zhaga.

Effective compliance with lighting standards and regulations requires a thorough understanding of the requirements, as well as the applicable technologies, tools, and testing methods. Lighting professionals should also keep abreast of the latest updates and revisions to the standards and regulations, in order to ensure that their designs and installations are up-to-date and compliant.

Performance standards refer to guidelines that specify the minimum acceptable levels of performance for lighting systems. These standards aim to ensure that lighting systems are energy-efficient, safe, and provide adequate illumination for their intended use. Some examples of performance standards related to lighting systems include the following:

- Illuminating Engineering Society of North America (IESNA) standards: The IESNA develops and publishes standards and guidelines for lighting systems in various applications, such as commercial buildings, outdoor lighting, and sports lighting. These standards cover various aspects of lighting design, including illuminance levels, uniformity, color quality, and energy efficiency.
- National Electrical Code (NEC): The NEC is a model electrical code developed by the National Fire Protection Association that covers the installation of electrical systems in buildings. The code includes requirements related to lighting systems, such as wiring methods, grounding, and overcurrent protection.
- International Electrotechnical Commission (IEC) standards: The IEC develops and publishes international standards for electrical and electronic technologies, including lighting systems. These standards cover various aspects of lighting systems, such as performance, safety, and electromagnetic compatibility.
- Energy codes and standards: Energy codes and standards, such as ASHRAE 90.1 and the International Energy Conservation Code (IECC), include performance requirements related to lighting systems, such as lighting power density limits and automatic controls. These codes aim to promote energy-efficient lighting design and reduce energy consumption [27–34].

References

1. Halle, M., & Meng, J. (2003). Lightkit: A lighting system for effective visualization. In *IEEE visualization, 2003. VIS 2003* (pp. 363–370). IEEE.
2. Muhamad, W. N. W., Zain, M. Y. M., Wahab, N., Aziz, N. H. A., & Abd Kadir, R. (2010). Energy efficient lighting system design for building. In *2010 International conference on intelligent systems, modelling and simulation* (pp. 282–286). IEEE.
3. Soori, P. K., & Vishwas, M. (2013). Lighting control strategy for energy efficient office lighting system design. *Energy and Buildings, 66*, 329–337.
4. DiLaura, D. L. (2010). A new lighting handbook. *LEUKOS, 6*(4), 256–258.
5. Lechner, N. (2014). *Heating, cooling, lighting: Sustainable design methods for architects*. John Wiley & Sons.
6. Steidle, A., Werth, L., de Boer, J., & Sedlbauer, K. (2014). How to create sustainable lighting for users? Psychological mechanisms underlying lighting effects. In *Proceedings experiencing light 2014: International conference on the effects of light on wellbeing* (pp. 78–81). Technische Universiteit Eindhoven.
7. Zaeva-Burdonskaya, E. A., & Nazarov, Y. V. (2021). Lighting design: Facets of a single phenomenon. *Journal of Physics: Conference Series, 1828*(1), 012132. IOP Publishing.
8. Simpson, R. (2013). *Lighting control: Technology and applications*. Routledge.
9. Sabourin, N. T. (2017). *The effects of connected lighting on lighting controls and design*. Kansas State University.
10. DiLouie, C. (2021). *Advanced lighting controls: Energy savings, productivity, technology and applications*. CRC Press.
11. International Energy Conservation Code. (2018). International Code Council.
12. International Code Council. (2018). International Energy Conservation Code (IECC) toolkit.
13. U.S. Department of Energy. (2017). Compliance and enforcement of energy codes.
14. National Fire Protection Association. (2020). Codes and standards.
15. International Code Council. (2018). International Building Code (IBC) toolkit.
16. Canadian Standards Association. (2018). Canadian Electrical Code.
17. National Institute of Standards and Technology. (2017). Economic benefits of adopting the International Building Code.
18. United Nations. (2015). Safety standards for construction and operation of buildings.
19. United States Access Board. (2020). The ADA accessibility standards.
20. Accessibility Directorate of Ontario. (2020). Accessibility laws and standards.
21. UK Government. (2010). Equality Act 2010 guidance.
22. World Health Organization. (2011). World report on disability.
23. U.S. Green Building Council. (2021). LEED v4.1. https://www.usgbc.org/leed-v41
24. U.S. Environmental Protection Agency. (2021). Energy Star. https://www.energystar.gov/
25. International Code Council. (2021). International Energy Conservation Code. https://www.iccsafe.org/codes-tech-support/codes/international-energy-conservation-code/
26. California Energy Commission. (2021). Building energy efficiency standards. https://www.energy.ca.gov/title24/
27. Illuminating Engineering Society of North America. (2021). Standards and guidelines. https://www.ies.org/standards/
28. National Fire Protection Association. (2021). National Electrical Code. https://www.nfpa.org/codes-and-standards/all-codes-and-standards/list-of-codes-and-standards/detail?code=70.
29. International Electrotechnical Commission. (2021). Standards. https://www.iec.ch/standards/
30. U.S. Department of Energy. (2021). Building Energy Codes Program. https://www.energycodes.gov/

31. CIE. (2019). *CIE position statement on outdoor lighting and the environment*. International Commission on Illumination.
32. IEC. (2018). *IEC 60364: Electrical installations for buildings*. International Electrotechnical Commission.
33. ASHRAE. (2019). *Energy standard for buildings except low-rise residential buildings (ASHRAE 90.1)*. American Society of Heating, Refrigerating and Air-Conditioning Engineers.
34. ADA. (2010). *2010 ADA standards for accessible design*. US Department of Justice.

Chapter 5
Lighting Sources Used for Interior Design

Contents

5.1 Artificial Lighting Sources

Lighting sources are devices that produce visible light for the purpose of illuminating spaces or objects. There are various types of lighting sources, each with its own characteristics, benefits, and drawbacks. Here are some of the most common lighting sources:

G. Ozenen, *Architectural Interior Lighting*,
https://doi.org/10.1007/978-3-031-49695-0_5

Incandescent bulbs These bulbs use a filament that is heated by an electric current until it emits visible light. They are relatively cheap and easy to use, but they are not very energy-efficient and have a short lifespan.

5.1.1 Halogen Bulbs

These bulbs use a tungsten filament enclosed in a small quartz capsule filled with halogen gas. They are similar to incandescent bulbs in terms of their appearance and operation, but they are more efficient and have a longer lifespan.

Halogen bulbs are a type of incandescent lamp that use halogen gas to improve their efficiency and longevity. Here are some details about halogen bulbs, along with academic references:

- Operation: Halogen bulbs work by passing an electric current through a tungsten filament that is enclosed in a small quartz capsule filled with halogen gas. The halogen gas reacts with the tungsten atoms that have evaporated from the filament and redeposits them back onto the filament, extending its lifespan and maintaining a high light output.
- Efficiency: Halogen bulbs are more efficient than traditional incandescent bulbs because the halogen gas allows them to operate at higher temperatures and produce more light for the same amount of power. They can also provide a brighter and whiter light compared to incandescent bulbs.
- Applications: Halogen bulbs are commonly used in residential, commercial, and industrial lighting applications, such as track lighting, recessed lighting, and outdoor lighting. They are also used in automotive headlights, film and photography, and medical equipment.
- Environmental impact: Halogen bulbs contain small amounts of halogen gas and other hazardous materials that can be harmful to the environment if not disposed of properly. They also consume more energy compared to LED and CFL bulbs, making them less environmentally friendly in the long term [1].

5.1.2 Fluorescent Tubes

These tubes use an electric current to excite a gas inside a glass tube, which produces ultraviolet light that is converted into visible light by a phosphor coating on the tube. They are more energy-efficient than incandescent bulbs, but they can be bulky and have a flicker that some people find annoying.

Fluorescent tubes are a type of lamp that uses an electric current to excite mercury vapor and produce ultraviolet light. This light then interacts with a phosphor coating on the inside of the tube, producing visible light. Here are some details about fluorescent tubes, along with academic references:

- Efficiency: Fluorescent tubes are more efficient than traditional incandescent bulbs because they produce more light for the same amount of power. They can also last longer than incandescent bulbs, reducing the need for frequent replacements.
- Applications: Fluorescent tubes are commonly used in commercial and industrial settings, such as offices, warehouses, and schools. They are also used in residential applications, such as kitchens and garages, as well as in specialized applications, such as aquarium lighting and plant growth lighting.
- Color temperature and CRI: Fluorescent tubes are available in a range of color temperatures and color rendering indices (CRI), making them suitable for a variety of applications. They can produce a range of colors from warm white to cool white, and their CRI can vary from low to high, affecting the quality of the light they produce.
- Environmental impact: Fluorescent tubes contain small amounts of mercury, which can be harmful to the environment if not disposed of properly. They also consume more energy compared to LED and CFL bulbs, making them less environmentally friendly in the long term [2].

5.1.3 Compact Fluorescent Bulbs

These bulbs use the same technology as fluorescent tubes, but in a smaller and more compact form. They are more energy-efficient than incandescent bulbs and have a longer lifespan, but they can also have a flicker and may contain hazardous materials.

Compact fluorescent bulbs (CFLs) are a type of energy-efficient lighting that use a mix of gas and mercury vapor to produce ultraviolet light, which then interacts with a phosphor coating to produce visible light. Here are some details about CFLs, along with academic references:

- Efficiency: CFLs are more efficient than traditional incandescent bulbs because they use less energy to produce the same amount of light. They also have a longer lifespan than incandescent bulbs, reducing the need for frequent replacements.
- Applications: CFLs are commonly used in residential and commercial applications, such as homes, offices, and retail stores. They are available in a range of color temperatures and CRI values, making them suitable for a variety of lighting needs.
- Environmental impact: CFLs contain a small amount of mercury, which can be harmful to the environment if not disposed of properly. However, they are still considered more environmentally friendly than incandescent bulbs due to their energy efficiency and longer lifespan.
- Disadvantages: CFLs have some drawbacks compared to other types of energy-efficient lighting, such as LED bulbs. They can take a few moments to reach their full brightness, and their lifespan can be reduced if they are turned on and off frequently [3].

5.1.4 Light-Emitting Diodes (LEDs)

These devices use a semiconductor material that emits light when an electric current is applied. They are highly energy efficient, have a long lifespan, and are available in a wide range of colors and intensities. They are also becoming increasingly affordable and versatile, making them a popular choice for many lighting applications.

High-intensity discharge (HID) lamps: These lamps use an electric arc to produce light inside a sealed glass bulb. They are very bright and efficient, but they can be expensive and require a special ballast to operate.

There are also many specialty lighting sources, such as fiber optics, neon lights, and laser diodes, which are used for specific applications or effects. The choice of lighting source depends on various factors, such as the desired lighting quality, energy efficiency, color rendering, and cost-effectiveness.

Light-emitting diodes (LEDs) are a type of energy-efficient lighting that use semiconductor materials to produce light. Here are some details about LEDs, along with academic references:

- Efficiency: LEDs are the most energy-efficient lighting option available, using up to 80% less energy than traditional incandescent bulbs. They also have a longer lifespan than other types of lighting, reducing the need for frequent replacements.
- Applications: LEDs are commonly used in residential and commercial applications, such as homes, offices, and outdoor lighting. They are available in a range of color temperatures and CRI values, making them suitable for a variety of lighting needs.
- Environmental impact: LEDs are more environmentally friendly than other types of lighting due to their energy efficiency and longer lifespan. They also do not contain hazardous materials, such as mercury.
- Advantages: LEDs have several advantages compared to other types of lighting, including their energy efficiency, long lifespan, and durability. They also do not emit heat, making them safer to use [4].

5.2 Use of Artificial Lighting Sources in Interior Design

Artificial lighting sources play a crucial role in interior design, as they can create different moods and atmospheres, highlight features of the space, and enhance the functionality of the environment. The following are some ways in which artificial lighting sources can be used in interior design.

5.2.1 General Lighting

General lighting is the primary source of light in a room and provides overall illumination. It can be achieved through the use of ceiling-mounted fixtures, recessed lighting, or track lighting. General lighting should be bright enough to provide adequate illumination for the entire room. General lighting is the foundational lighting in a space, providing overall illumination to ensure that the room is well-lit and that all areas are visible. It is typically achieved through a combination of ceiling-mounted fixtures, such as recessed lights, surface-mounted fixtures, or pendant lights, and sometimes supplemented with wall-mounted fixtures or floor lamps. General lighting is essential in interior design as it sets the tone for a room and establishes the basic level of illumination.

The Illuminating Engineering Society (IES) provides guidelines for lighting levels for various applications in its publication "The Lighting Handbook," which is widely referenced in the lighting industry. According to the IES, recommended lighting levels for general lighting in interior spaces depend on the specific type of space and the tasks that are typically performed there. For example, a residential living room might require 10–20 foot-candles (fc) of light, while an office space might require 30–50 fc.

In addition to providing adequate illumination for a space, general lighting can also contribute to the aesthetic appeal of a room. By selecting the appropriate fixtures and light sources, designers can create a desired mood or ambiance, such as warm and cozy or bright and energizing. This can be achieved through the color temperature and color rendering of the light sources used, as well as through the design and placement of the fixtures themselves.

Overall, the use of general lighting in interior design is essential to provide adequate illumination for the tasks performed in a space while also contributing to the overall aesthetic and ambiance of the room [5–7].

5.2.2 Task Lighting

Task lighting is a specific type of lighting that is designed to illuminate a particular area or task, such as reading, cooking, or working on a computer. Task lighting is important in interior design as it helps to improve visual acuity, reduce eyestrain, and increase productivity.

Task lighting fixtures can take many forms, including desk lamps, floor lamps, and under-cabinet lighting. The type of task lighting used depends on the nature of the task and the desired level of illumination. Task lighting is used to illuminate specific areas where tasks are performed, such as reading, cooking, or working. It can be achieved through the use of desk lamps, table lamps, or under-cabinet lighting.

For example, in a kitchen, under-cabinet lighting can be used to provide task lighting for food preparation and cooking, while a desk lamp can be used to provide task lighting for reading or working on a computer. In an office setting, task lighting can be used to provide illumination for individual workstations, while in a classroom, task lighting can be used to illuminate the whiteboard or individual student work areas.

Proper task lighting is important for reducing eye strain and improving productivity and can also contribute to energy efficiency by allowing for the use of lower levels of ambient lighting [8].

5.2.3 Accent Lighting

Accent lighting is a type of lighting used in interior design to highlight or draw attention to specific objects, areas, or features of a space. It is often used to create a focal point or to add visual interest to a room. Accent lighting can be achieved using a variety of lighting fixtures, such as spotlights, track lighting, or wall sconces. Accent lighting is used to highlight specific features of a room, such as artwork, architectural details, or decorative objects. It can be achieved through the use of track lighting, wall-mounted fixtures, or recessed lighting.

One common use of accent lighting is to highlight artwork or other decorative objects in a room. By directing a spotlight or other fixture onto the object, it can be showcased and given greater prominence within the space. Accent lighting can also be used to highlight architectural features, such as columns or arches, or to draw attention to a particular area of a room, such as a seating area or dining table.

In addition to its aesthetic benefits, accent lighting can also provide practical benefits in interior design. For example, it can be used to provide additional task lighting in areas where it is needed, such as in a home office or kitchen [9].

5.2.4 Decorative Lighting

Decorative lighting is used to enhance the aesthetic appeal of a space and to create a mood or atmosphere. It includes fixtures such as chandeliers, wall sconces, and table lamps, which are often chosen for their decorative qualities rather than their functional purposes. Decorative lighting can be used in combination with other types of lighting to create layered lighting schemes.

In interior design, decorative lighting is often used as a focal point or statement piece in a room. For example, a large chandelier can be used to draw attention to a dining table or a grand foyer, while a series of small pendants can be used to add interest to a hallway or staircase. Decorative lighting can also be used to highlight artwork or architectural features.

When designing with decorative lighting, it is important to consider the overall style and theme of the space. The fixtures should complement the design elements and create a cohesive look. Additionally, the lighting should be chosen based on the size of the space and the desired level of illumination [10].

5.3 Natural Lighting Sources

Natural lighting sources refer to the light that comes from the sun or the sky. These sources of light are uncontrolled and can vary depending on the time of day, season, weather, and location. The common natural lighting sources are direct sunlight, diffuse skylight, reflected light, and daylight systems [11].

5.3.1 Direct Sunlight

Direct sunlight is the brightest and most intense source of natural light. It can create sharp shadows and cause glare, but it also provides high color rendering and can enhance the visual appeal of a space.

Direct sunlight refers to the sunlight that is not obstructed by clouds or other objects and is the most intense form of natural light. Here are some details about direct sunlight, along with academic references:

- Intensity: Direct sunlight is the most intense form of natural light, and its intensity varies depending on factors such as time of day, season, and latitude.
- Color temperature: Direct sunlight has a color temperature of around 5500 K, which is considered a "cool" color temperature and produces a bluish-white light.
- Health effects: Exposure to direct sunlight can have both positive and negative health effects. It is the primary source of vitamin D, which is essential for bone health, but overexposure can lead to skin damage and an increased risk of skin cancer.
- Design considerations: Direct sunlight can be used in building design to provide natural lighting and reduce the need for artificial lighting. However, it can also cause glare and heat gain, so proper shading and orientation of windows is important [12].

5.3.2 Diffuse Skylight

This is the light that is scattered by the atmosphere and enters a space through windows, skylights, or openings. It provides a more even and soft illumination, but it can also be less bright and have a cooler color temperature.

Diffuse skylight is a form of natural light that is scattered and dispersed by the atmosphere, creating a soft, even light that does not produce harsh shadows. Here are some details about diffuse skylight, along with academic references:

- Characteristics: Diffuse skylight is characterized by its evenness and lack of directionality. It is the result of sunlight being scattered by the atmosphere, with shorter wavelengths (blue light) being scattered more than longer wavelengths (red light).
- Color temperature: The color temperature of diffuse skylight varies depending on the time of day and weather conditions. On a clear day, it can range from around 5000 K to 6500 K, while on a cloudy day, it may be lower.
- Health effects: Exposure to diffuse skylight can have positive effects on mood and productivity, as well as circadian rhythm regulation. It is also a source of vitamin D, although to a lesser extent than direct sunlight.
- Design considerations: Diffuse skylight can be used in building design to provide natural lighting and reduce the need for artificial lighting. It can also be used to create a sense of openness and connection to the outdoors. However, it can cause glare and heat gain, so proper shading and orientation of windows is important [12].

5.3.3 Reflected Light

Reflected light is the light that is bounced off surfaces such as walls, floors, or ceilings. It can add a subtle and indirect illumination to a space, but it can also cause uneven lighting or color casts depending on the reflectance properties of the surface.

Reflected light is the light that bounces off of a surface and enters our eyes. It plays an important role in our perception of color, as the colors we see are the result of the wavelengths of light that are reflected back to us. Here are some details about reflected light, along with academic references:

- Characteristics: Reflected light can take on different characteristics depending on the surface it is bouncing off of. A smooth, shiny surface will reflect light more evenly and create more specular highlights, while a rough or matte surface will scatter the light and create more diffuse reflections.
- Color rendition: Reflected light can affect the color rendition of an object, as it can alter the color temperature and intensity of the light. This can be a concern in color critical applications such as art conservation, where accurate color reproduction is important.
- Design considerations: Reflected light can be used in lighting design to enhance the perception of space and create a sense of depth. It can also be used to highlight certain features or objects in a space. However, it can also cause glare and visual discomfort, so careful placement of light sources and use of appropriate materials is important.

- Measurement: Reflected light can be measured using a variety of instruments, including spectrophotometers and colorimeters. These instruments can provide information on the color and intensity of the reflected light, as well as its distribution across the surface [13].

5.3.4 Daylighting Systems

Daylighting systems are passive or active strategies that use natural light to provide illumination in buildings. Examples include light shelves, clerestories, and skylights with light-diffusing elements. These systems can enhance energy efficiency, occupant comfort, and visual comfort in a space.

Natural lighting sources have many benefits for human health and well-being, including circadian rhythm regulation, vitamin D synthesis, and mood enhancement. They are also a sustainable and renewable source of energy, reducing the reliance on artificial lighting and electricity. However, they can also have drawbacks, such as heat gain, glare, and fading of materials. Therefore, a proper design and management of natural lighting sources is important to maximize their benefits and minimize their drawbacks.

Daylighting systems are designed to bring natural light into buildings and reduce the need for artificial lighting. Some details about daylighting systems are as follows:

- Types of systems: There are several types of daylighting systems, including skylights, light tubes, and light shelves. Skylights are windows installed in the roof, while light tubes are tubes that reflect and direct natural light into interior spaces. Light shelves are horizontal surfaces that reflect and redirect sunlight into a space.
- Benefits: Daylighting systems can provide numerous benefits, including energy savings, improved occupant comfort and well-being, and increased productivity. They can also enhance the aesthetic quality of a space and create a connection to the outdoors.
- Design considerations: Designing a successful daylighting system requires careful consideration of factors such as building orientation, glazing properties, shading, and control systems. The goal is to optimize the amount of natural light while minimizing glare and heat gain.
- Measurement: Daylighting performance can be measured using a variety of metrics, including daylight factor, spatial daylight autonomy, and annual sunlight exposure. These metrics can help designers and building operators evaluate the effectiveness of their daylighting systems and make adjustments as needed [14].

5.4 Use of Natural Lighting Sources in Interior Design

Natural lighting sources can be an excellent addition to an interior design scheme, providing a range of benefits from energy savings to enhancing mood and well-being. Here are some ways natural lighting sources can be utilized in interior design:

5.4.1 Maximizing Daylight

One of the most common ways to utilize natural lighting sources in interior design is to maximize the amount of natural daylight entering the space. This can be achieved through careful positioning of windows and skylights, as well as through the use of light shelves, reflectors, and other daylighting strategies.

Maximizing daylight use is a common approach to improving the energy efficiency and occupant comfort of buildings. Here are some ways that daylight can be maximized in interior design:

- Window placement and size: Properly positioning and sizing windows is key to maximizing natural light in a space. South-facing windows receive the most direct sunlight throughout the day, while east- and west-facing windows receive sunlight during specific times of day. North-facing windows receive the least amount of sunlight. Proper window placement can help to distribute natural light evenly throughout a space.
- Glazing: The type of glass used in windows can affect the amount of natural light that enters a space. High-performance glazing can allow more sunlight into a space while reducing heat gain and glare. Additionally, adding window film or shading devices can help to control the amount of light and heat that enters a space.
- Daylight harvesting: Daylight harvesting is the practice of using automated systems to control artificial lighting based on the amount of natural light present in a space. This can help reduce energy consumption and improve occupant comfort. Photo sensors or light sensors can be used to adjust lighting levels based on the amount of daylight present.
- Interior design: Proper interior design can help maximize the benefits of natural light. Light-colored finishes and reflective surfaces can help distribute light evenly throughout a space. Additionally, using interior partitions or translucent materials can help filter and diffuse natural light [15].

5.4.2 Daylight Harvesting

Daylight harvesting involves using sensors and other devices to automatically adjust artificial lighting levels based on the amount of natural light available. This can help reduce energy consumption and improve comfort levels.

Daylight harvesting is a technique used in interior design to make maximum use of natural light for lighting and energy savings. It involves designing a building and its lighting system to take advantage of daylight and control electric lighting automatically to provide appropriate light levels. This technique can help reduce energy consumption and lower electricity bills while improving the visual environment of the space.

There are various methods to implement daylight harvesting in interior design. One common approach is to use sensors to detect the amount of natural light available in the space and adjust the artificial lighting accordingly. This can include dimming or turning off electric lights when there is sufficient natural light available, or increasing the intensity of the artificial lighting when natural light is limited.

Daylight harvesting can also involve designing a building to optimize natural light, such as using skylights, light shelves, or reflective surfaces to direct and distribute natural light into the space. The use of glazing materials and window placement can also be optimized to take advantage of natural light while minimizing glare and heat gain.

Research studies have shown that daylight harvesting can lead to significant energy savings in commercial buildings, particularly in spaces that are typically occupied during daylight hours, such as offices, schools, and retail spaces.

Overall, daylight harvesting is a valuable technique in interior design for promoting sustainability and energy efficiency while improving the visual and thermal comfort of the space [16].

5.4.3 Biophilic Design

Biophilic design is an approach that incorporates natural elements and materials into interior spaces to improve well-being and productivity. This can include the use of natural lighting sources such as sunlight, which has been shown to enhance mood and cognitive performance.

Biophilic design in interior design is the concept of incorporating natural elements, such as plants, water, and sunlight, into the built environment to improve occupants' physical and psychological well-being. It is based on the idea that humans have an innate connection to nature and that being in contact with natural elements can reduce stress, enhance cognitive function, and increase creativity.

The use of biophilic design elements in interior design can range from simple additions such as potted plants to larger-scale interventions such as green walls and skylights. Biophilic design can also be integrated into the overall design concept, such as the use of natural materials and colors [17].

5.4.4 Solar Shading

Solar shading devices such as shades, blinds, and curtains can be used to control the amount of natural light entering a space, helping to reduce glare and heat gain while still allowing for daylight to enter.

Solar shading is a design strategy used to control the amount of solar heat gain and glare in a space while still allowing natural light to enter. This is typically achieved through the use of shading devices such as overhangs, fins, louvers, or awnings, which are strategically placed to block direct sunlight from entering the building.

The use of solar shading devices can improve the energy efficiency of a building by reducing the amount of cooling required to maintain a comfortable indoor temperature, which can lead to significant cost savings. Additionally, solar shading can improve occupant comfort by reducing glare and improving visual comfort.

Research has shown that the use of solar shading can have a significant impact on building energy performance. The use of external shading devices was found to reduce the cooling load of a building, resulting in a significant reduction in energy consumption.

In addition to energy efficiency benefits, the use of solar shading can also contribute to the aesthetic and functional aspects of a building design. Shading devices can be designed to enhance the visual appeal of a building and can be used to create outdoor spaces that are protected from direct sunlight and glare.

Overall, the use of solar shading is an effective strategy for improving the energy efficiency and comfort of a building. Through careful design and selection of shading devices, architects and designers can create buildings that are more sustainable, efficient, and comfortable for occupants [18].

References

1. U.S. Environmental Protection Agency. (2019). Lighting facts: Halogen bulbs. Retrieved from https://www.energystar.gov/products/lighting_fans/light_bulbs/learn_about_led_bulbs
2. Heffernan, W. J. B., Frater, L. P., & Watson, N. R. (2007). LED replacement for fluorescent tube lighting. In *2007 Australasian universities power engineering conference* (pp. 1–6). IEEE.
3. Luther, L. (2008). *Compact fluorescent light bulbs (CFLs): Issues with use and disposal.* Congressional Research Service, the Library of Congress.
4. Schubert, E. F. (2018). *Light-emitting diodes (2018).* E. Fred Schubert.
5. Imamguluyev, R. (2021). Application of fuzzy logic model for correct lighting in computer aided interior design areas. In *Intelligent and fuzzy techniques: Smart and innovative solutions: Proceedings of the INFUS 2020 conference, Istanbul, Turkey, July 21–23, 2020* (pp. 1644–1651). Springer International Publishing.
6. Sandeva, V., & Despot, K. (2017). The effect of lighting exterior and interior design. *Innovation and Entrepreneurship, 5*(1), 10–21.
7. Ding, Y., & Liu, C. (2015). Design of light utilization model in interior design. In *International conference on education, management and computing technology (ICEMCT-15)* (pp. 1234–1237). Atlantis Press.

8. Newsham, G., Arsenault, C., Veitch, J., Tosco, A. M., & Duval, C. (2005). Task lighting effects on office worker satisfaction and performance, and energy efficiency. *LEUKOS, 1*(4), 7–26.
9. Trajkova, A., & Namichev, P. (2022). Impact of new lighting technologies as an important element of the interior. *Natural Resources and Technology, 16*(2), 48–55.
10. Tantanatewin, W., & Inkarojrit, V. (2016). Effects of color and lighting on retail impression and identity. *Journal of Environmental Psychology, 46*, 197–205.
11. Wlas, M., & Galla, S. (2018). The influence of LED lighting sources on the nature of power factor. *Energies, 11*(6), 1479.
12. Minnaert, M. (2013). *The nature of light and colour in the open air*. Courier Corporation.
13. Blinn, J. F. (1982). Light reflection functions for simulation of clouds and dusty surfaces. *ACM Siggraph Computer Graphics, 16*(3), 21–29.
14. Whang, A. J. W., Yang, T. H., Deng, Z. H., Chen, Y. Y., Tseng, W. C., & Chou, C. H. (2019). A review of daylighting system: For prototype systems performance and development. *Energies, 12*(15), 2863.
15. Widayat, R., & Studyanto, A. B. (2021). Utilization of natural resources in supporting eco-interior design. *IOP Conference Series: Earth and Environmental Science, 905*(1), 012034. IOP Publishing.
16. El Sheikh, M., & Kensek, K. (2011). Intelligent skins: Daylight harvesting through dynamic light-deflection in office spaces. In *En ARCC 2011 conference proceedings* (pp. 293–304). ARCC Conference Repository.
17. Richardson, M., & Butler, C. W. (2022). Nature connectedness and biophilic design. *Building Research & Information, 50*(1–2), 36–42.
18. Wright, J. L., Kotey, N. A., Barnaby, C. S., & Collins, M. R. (2009). Solar gain through windows with shading devices: Simulation versus measurement. *ASHRAE Transactions, 115*(2), 18.

Chapter 6
Decorative Lighting for Interior Design

Contents

Decorative lighting plays an important role in creating a visually appealing and aesthetically pleasing interior environment. It serves as an additional layer of lighting that enhances the visual impact of the space and highlights specific areas, features, or objects. Here are some details on the use of decorative lighting in interior design, supported by relevant references:

- Chandeliers: Chandeliers are often used as a focal point in the interior design scheme. They add an element of grandeur and elegance to the space and are available in a variety of styles and sizes to suit different design preferences.

G. Ozenen, *Architectural Interior Lighting*,
https://doi.org/10.1007/978-3-031-49695-0_6

Chandeliers can be made of a variety of materials, such as crystal, glass, metal, or wood.

- Pendant lights: Pendant lights are versatile and can be used in a variety of interior design styles. They are commonly used as accent lighting, particularly over kitchen islands or dining tables. Pendant lights come in a variety of shapes, sizes, and materials, including glass, metal, and fabric.
- Wall sconces: Wall sconces are a popular decorative lighting option for interior spaces. They are typically used to create a warm and inviting atmosphere in the room and can be used to highlight specific features, such as artwork or architectural details. Wall sconces are available in a range of styles and finishes to complement different design schemes.
- Table and floor lamps: Table and floor lamps are used to provide both ambient and task lighting in a space. They come in a variety of sizes, styles, and materials and can be used to create a cohesive design scheme in the room. Table and floor lamps can be used to provide a decorative accent to a space or to highlight specific areas such as a reading nook or workspace.
- LED strip lights: LED strip lights are a popular choice for decorative lighting as they are versatile and easy to install. They can be used to highlight specific areas, such as under cabinets, along shelves, or around architectural details. LED strip lights come in a range of colors and can be dimmed to create a desired ambiance.

Overall, decorative lighting plays an important role in enhancing the visual impact of an interior space. By selecting the right decorative lighting options, designers can create a cohesive and visually appealing design scheme that highlights the unique features and characteristics of the space [1].

6.1 Chanderliers

6.1.1 Use of Chandeliers in Interior Design

Chandeliers are decorative lighting fixtures that are often used in interior design to provide a statement piece and enhance the visual appeal of a space. They are characterized by a multi-armed design with hanging crystals or other decorative elements and are often installed in entryways, dining rooms, or other large open spaces.

In terms of design, chandeliers come in a variety of styles and materials, including traditional crystal, modern metal, or even natural materials such as wood or bamboo. They can be used to add a touch of elegance to a space or to create a bold and dramatic focal point.

When selecting a chandelier for an interior design project, it is important to consider the scale and proportion of the space, as well as the overall design style. Additionally, factors such as the height of the ceiling, the amount of natural light in the room, and the intended use of the space should be taken into account.

Some popular chandelier styles include the following:

- Traditional crystal chandeliers, which feature hanging glass or crystal prisms and are often associated with formal spaces such as dining rooms or ballrooms
- Contemporary or modern chandeliers, which feature clean lines and geometric shapes and are often made of metal or other industrial materials
- Rustic or natural chandeliers, which are made from natural materials such as wood or bamboo and are often used in more casual or eclectic spaces.

6.1.2 Types of Chandeliers in Interior Design

There are different types of chandeliers that vary in size, shape, materials, and style. Some of the most common types of chandeliers are as follows:

- Crystal Chandeliers: These are the most classic type of chandeliers and feature cut glass or crystal prisms that reflect light and create a dazzling display.
- Candle Chandeliers: As the name suggests, these chandeliers resemble candles and are often made from wrought iron or other metals. They are designed to look like traditional candle holders and provide a warm and inviting glow.
- Tiered Chandeliers: These chandeliers have multiple tiers of arms that extend from a central point, giving them a more elaborate and grand appearance.
- Contemporary Chandeliers: These chandeliers have a modern design and are made from materials like glass, metal, and plastic. They often have clean lines and a minimalistic look.
- Rustic Chandeliers: These chandeliers are made from natural materials like wood and feature a more rustic or country-style design.
- Beaded Chandeliers: These chandeliers feature strands of beads or crystals that dangle from the arms, creating a shimmering effect.
- Drum Chandeliers: These chandeliers have a drum-shaped shade that encircles the bulbs, creating a diffused and soft light.
- Empire Chandeliers: These chandeliers have a dome-shaped frame that is adorned with crystal or glass accents and has arms that extend outward.

These are just a few examples of the many types of chandeliers that are available. The choice of chandelier depends on the style and mood of the room, as well as the personal preferences of the designer or homeowner [2, 3].

6.2 Pendant Lights

6.2.1 Use of Pendant Lights in Interior Design

Pendant lights are a popular choice for interior lighting design due to their versatility and aesthetics. They can be used to provide both ambient and task lighting, as well as serve as a decorative feature. Here are some details about pendant lights in interior design with relevant references:

- Types of Pendant Lights: Pendant lights come in various styles, shapes, and sizes. Some popular types include dome pendants, drum pendants, multi-light pendants, and mini pendants. Each type has its unique characteristics that make it suitable for different interior design applications.
- Placement: Pendant lights are often used in kitchens, dining areas, and bedrooms. They can be installed over tables, islands, and countertops to provide task lighting, or in the center of the room to provide ambient lighting. The height and placement of the pendant light should be carefully considered to ensure proper illumination and avoid glare.
- Materials: Pendant lights can be made from a wide range of materials, including glass, metal, fabric, and even natural materials like wood and rattan. The material used can impact the overall aesthetics of the light fixture and the interior space it is placed in.
- Bulbs: Pendant lights can be used with various types of bulbs, including LED, incandescent, and halogen bulbs. The choice of bulb can impact the color temperature and brightness of the light emitted, as well as the energy efficiency of the fixture.
- Dimming: The ability to dim pendant lights can enhance their versatility and usefulness in interior design. Dimming controls can be incorporated into the light fixture itself or added separately to the wall or switch.

6.2.2 Types of Pendant Lights in Interior Design

There are different types of pendant lights available for use in interior design. Here are some examples:

1. Mini pendant lights: These are small pendant lights that are perfect for use in groups or clusters. They are often used for accent lighting or task lighting.
2. Drum pendant lights: These are pendant lights that have a drum-shaped shade. They can provide diffused lighting and are often used as a decorative element in a room.
3. Globe pendant lights: These are pendant lights that have a globe-shaped shade. They can provide diffused lighting and are often used as a decorative element in a room.
4. Cluster pendant lights: These are pendant lights that come in a cluster or grouping of lights. They can be used to create a dramatic effect in a room.
5. Linear pendant lights: These are pendant lights that have a long, narrow shape. They are often used to provide task lighting over a kitchen island or dining table.
6. Multi-light pendant lights: These are pendant lights that have multiple lights on a single fixture. They can be used to provide both general and task lighting in a room.

These are just a few examples of the different types of pendant lights available. The type of pendant light that is best for a particular space will depend on factors such as the size of the room, the style of the room, and the desired lighting effect [1, 4].

6.3 Wall Sconces

6.3.1 Use of Wall Sconces in Interior Design

Wall sconces are a popular form of decorative lighting that is often used in interior design. They can provide ambient or accent lighting, and they come in a variety of styles to match any decor. Wall sconces are typically installed on a wall, either singly or in pairs, and they can be used to create a variety of lighting effects.

One benefit of wall sconces is that they can be used to free up floor space, which can be especially useful in smaller rooms. They are also versatile and can be used in a variety of settings, from living rooms and dining rooms to hallways and bathrooms. In addition, wall sconces can be used to create a dramatic effect, particularly when used in combination with other lighting sources.

Wall sconces can be designed to use a variety of light sources, including incandescent, fluorescent, LED, and halogen bulbs. The choice of bulb will depend on the desired lighting effect and energy efficiency considerations.

One important consideration when installing wall sconces is the height at which they are placed on the wall. The American Lighting Association recommends that wall sconces be installed at eye level, which is typically between 60 and 66 inches from the floor.

6.3.2 Types of Wall Scones in Interior Design

There are different types of wall sconces that can be used in interior design. Some common types are as follows:

1. Up-light sconces: These sconces direct light upward, creating a soft, ambient glow that can help accentuate a room's architecture or decorative elements.
2. Down-light sconces: These sconces direct light downward, providing task lighting that can be useful in areas such as hallways, bathrooms, or reading nooks.
3. Up-down sconces: These sconces direct light both upward and downward, providing both ambient and task lighting in a single fixture.
4. Swing-arm sconces: These sconces feature an adjustable arm that can be moved to direct light where it is needed, making them useful as task lighting in areas such as home offices or bedrooms.
5. Picture lights: These sconces are designed to highlight artwork or other decorative elements on walls, providing a focused beam of light that can help to draw attention to the object being illuminated.

6. Candle sconces: These sconces are designed to resemble candle holders, providing a warm, traditional aesthetic that can be particularly effective in classic or rustic interior design schemes.

There are also many different styles of wall sconces available, ranging from modern and minimalist to ornate and traditional. The choice of sconce style will depend on the overall design goals of the space, as well as the specific needs for lighting in the area where the sconces will be installed [3].

6.4 Table and Floor Lamps

6.4.1 Use of Table and Floor Lambs in Interior Design

Table and floor lamps are commonly used as portable lighting sources in interior design. These fixtures can be moved around and positioned in various locations to provide task lighting or ambient lighting as needed. Here are some details about table and floor lamps used in interior design with references:

- Task Lighting: Table lamps and floor lamps are often used to provide task lighting for reading, writing, or other activities that require focused illumination. The height and placement of the lamp can be adjusted to direct light where it is needed.
- Ambient Lighting: Table lamps and floor lamps can also be used to provide ambient lighting in a room. The color temperature and brightness of the bulb can be selected to create the desired mood or atmosphere.
- Decorative Purposes: In addition to providing light, table lamps and floor lamps can also serve as decorative accents in a room. They come in a variety of styles, sizes, and materials, allowing them to complement or contrast with other design elements in the space.
- Design Considerations: When selecting a table or floor lamp for interior design, factors such as the size and scale of the room, the color scheme, and the intended use of the lamp should be considered. Additionally, the type of bulb used in the fixture can impact the quality and color of the light emitted.

6.4.2 Types of Table and Floor Lambs in Interior Design

There are different types of table and floor lamps, each with its own design, features, and functionality. Some common types of table and floor lamps include the following:

- Desk Lamps: Desk lamps are designed to provide task lighting for workspaces such as desks, reading tables, and nightstands. They usually have a flexible arm or a pivoting shade that allows you to direct the light exactly where you need it.

- Reading Lamps: Similar to desk lamps, reading lamps are designed to provide bright, focused light for reading. They typically have a tall, thin stem and a small, adjustable shade that can be positioned to illuminate a book or a magazine.
- Arc Lamps: Arc lamps feature a long, curved arm that extends outward from the base and arcs over the seating area. They are often used in living rooms and lounges to provide overhead lighting without taking up too much space.
- Torchiere Lamps: Torchiere lamps feature a tall, narrow stem that widens at the top to create a bowl-shaped shade that directs light upward. They are often used in living rooms and bedrooms to provide indirect lighting and create a warm, cozy atmosphere.
- Tripod Lamps: Tripod lamps feature a three-legged base that adds a unique and modern touch to any room. They come in a variety of styles and finishes, including wood, metal, and glass.
- Floor Lamps with Shelves: These types of lamps combine a floor lamp with storage shelves, providing both functional and ambient lighting as well as extra space for books, magazines, and decorative items [3, 5].

6.5 LED Strip Lights

6.5.1 Use of LED Strip Lights in Interior Design

LED strip lights are a popular lighting solution for adding ambient and accent lighting to interior spaces. They consist of a flexible circuit board with surface-mounted light-emitting diodes (LEDs) arranged in a series, which can be cut to custom lengths and installed in a variety of locations.

LED strip lights have several advantages over traditional lighting sources, including high energy efficiency, low heat output, and long lifespan. They also offer a range of color temperatures and color-changing options, making them versatile for different design applications.

In interior design, LED strip lights can be used in a variety of ways:

1. Cove lighting: Placing LED strip lights in a recessed area above cabinetry or along the perimeter of a room can create a soft and indirect lighting effect that highlights architectural details and adds a warm glow to the space.
2. Under cabinet lighting: Installing LED strip lights underneath kitchen or bathroom cabinets can provide task lighting for countertop activities while adding visual interest and depth to the space.
3. Accent lighting: Using LED strip lights to highlight artwork, architectural elements, or decorative objects can create a dramatic effect and draw attention to specific features in a room.
4. Backlighting: Placing LED strip lights behind TVs, headboards, or shelving can create a dynamic and inviting atmosphere in a space while also providing functional lighting.

5. Stair lighting: Installing LED strip lights on stair risers or handrails can create a safe and dramatic effect that adds depth and dimension to the space.

6.5.2 Types of LED Strip Lights in Interior Design

LED strip lights are not considered as table and floor lamps, but rather a separate lighting fixture. However, there are different types of LED strip lights that can be used in interior design, including the following:

- Flexible LED strip lights: These are the most commonly used LED strip lights, consisting of a flexible circuit board with LED chips mounted on it. They can be cut to size and easily installed on curved surfaces, making them ideal for use in a variety of applications such as under-cabinet lighting, cove lighting, and accent lighting.
- Rigid LED strip lights: These are similar to flexible LED strip lights, but have a rigid backing that allows them to be mounted on flat surfaces. They are often used for architectural lighting, such as highlighting the edges of steps and walkways.
- High-CRI LED strip lights: These LED strip lights have a Color Rendering Index (CRI) of 90 or higher, which means they are able to accurately reproduce colors. They are ideal for use in applications where color accuracy is important, such as in art galleries, museums, and retail spaces.
- RGB LED strip lights: These LED strip lights contain red, green, and blue LEDs, which can be mixed to create a wide range of colors. They are often used for decorative lighting, such as in restaurants, bars, and nightclubs.
- Tunable white LED strip lights: These LED strip lights can be adjusted to produce a range of color temperatures, from warm white to cool white. They are often used in applications where the color of the light needs to be changed depending on the time of day or the activity being performed, such as in offices and schools [6, 7].

6.6 Dimmers

Dimmers are devices that allow users to adjust the intensity of artificial lighting sources in a space, providing control over the lighting environment. In interior design, dimmers are commonly used to create ambiance and mood, as well as to increase energy efficiency by reducing the amount of energy used for lighting.

There are different types of dimmers available, including rotary, slide, and touch dimmers, and they can be used with various types of light sources, including incandescent, halogen, fluorescent, and LED lights.

Overall, dimmers are an important tool in interior design for providing flexibility and control over the lighting environment, as well as contributing to energy efficiency and occupant well-being.

Dimmers are a popular and effective way to adjust the brightness of decorative lighting elements in interior design. There are several types of dimmers available, each with its own advantages and disadvantages.

- Incandescent/halogen dimmers: These are the most common type of dimmer and are designed to work with incandescent and halogen light bulbs. They use a simple triac circuit to control the amount of voltage that is delivered to the bulb. They are relatively inexpensive and work well with most decorative lighting fixtures.
- Magnetic low-voltage dimmers: These dimmers are designed to work with low-voltage lighting systems that use magnetic transformers. They are more expensive than incandescent/halogen dimmers, but they offer better control over the brightness of the lights and can be used with a wider range of lighting fixtures.
- Electronic low-voltage dimmers: These dimmers are designed to work with low-voltage lighting systems that use electronic transformers. They are more expensive than magnetic low-voltage dimmers, but they offer smoother and more precise control over the brightness of the lights.
- LED dimmers: These dimmers are specifically designed to work with LED lighting systems. They use a different type of circuit than incandescent/halogen dimmers, as LED bulbs require a constant current rather than a constant voltage. LED dimmers are more expensive than other types of dimmers, but they are essential for ensuring that LED lights perform well and last a long time.

In addition to these types of dimmers, there are also wireless dimming systems that can be controlled using a smartphone or other device. These systems are more expensive than traditional dimmers, but they offer greater convenience and flexibility.

Overall, the type of dimmer that is best for a particular decorative lighting element will depend on the type of bulb and transformer that is being used, as well as the desired level of control and flexibility. It is important to consult with a lighting professional or electrician to determine the best dimming solution for your specific needs [3, 8].

6.6.1 Incandescent/Halogen Dimmers

Incandescent and halogen dimmers are some of the most commonly used dimming controls in decorative lighting design. These dimmers work by varying the voltage supplied to the light source, thereby changing the light output. They are often used with incandescent and halogen bulbs due to their compatibility with the resistive loads of these types of bulbs.

Incandescent/halogen dimmers come in a variety of types, including rotary, slide, and touch dimmers. Rotary dimmers are the most traditional and are operated by turning a knob to adjust the brightness of the light. Slide dimmers are operated by sliding a switch up or down to adjust the brightness. Touch dimmers, on the other hand, use a touch sensor to adjust the brightness with a simple touch.

When using incandescent/halogen dimmers in decorative lighting design, it is important to consider the compatibility of the dimmer with the type of bulb being used. Certain types of bulbs, such as low-voltage halogen bulbs, may require a specific type of dimmer to function properly [8].

6.6.2 Magnetic Low-Voltage Dimmers

Magnetic low-voltage (MLV) dimmers are another type of dimmer commonly used in decorative lighting design. These dimmers work by regulating the voltage supplied to a magnetic transformer that powers the lighting source. As the voltage is lowered, the magnetic field within the transformer is weakened, reducing the output voltage and therefore the light intensity.

MLV dimmers are particularly suited for use with low-voltage lighting systems, such as those that use halogen bulbs or LED tape, as they provide smooth and flicker-free dimming. They are also compatible with a wider range of lighting loads than incandescent/halogen dimmers, making them a more versatile option for decorative lighting design.

One consideration when using MLV dimmers is their compatibility with the specific lighting load being used. Some LED lighting systems may not be compatible with MLV dimmers, as the transformer may not work properly with the lower voltage output. It is important to carefully choose a dimmer that is compatible with the specific lighting load being used to ensure proper functionality and avoid damage to the lighting system [9].

6.6.3 Electronic Low-Voltage Dimmers

Electronic low-voltage (ELV) dimmers are another type of dimmer commonly used in decorative lighting elements. These dimmers are designed specifically to work with electronic transformers, which are often used to power low-voltage lighting systems such as LED strip lights and some types of pendant lights.

ELV dimmers use electronic circuitry to regulate the amount of voltage delivered to the lighting system, allowing users to adjust the brightness of their lights. These dimmers are typically more expensive than incandescent/halogen dimmers but are more efficient and offer smoother dimming performance.

ELV dimmers offer an efficient and effective way to control the brightness of decorative lighting elements in interior design while reducing energy consumption and improving lighting quality [9, 10].

6.6.4 LED Dimmers

LED dimmers are an essential element of modern interior design, allowing users to adjust the brightness of LED lighting systems and customize the mood and ambiance of a space. LED dimmers work by controlling the amount of current that passes through an LED light source, which affects its brightness level. There are several types of LED dimmers available in the market, each with unique features and performance characteristics.

One of the most common types of LED dimmers is the pulse-width modulation (PWM) dimmer. PWM dimmers work by rapidly switching the power supply to the LED light source on and off, which effectively reduces the average power supplied to the light and lowers its brightness. Another type of LED dimmer is the constant-current driver (CCD) dimmer, which uses a microcontroller to regulate the current flowing through the LED and adjust its brightness.

In addition to PWM and CCD dimmers, there are also many other types of LED dimmers, including resistive dimmers, phase-cut dimmers, and digital dimmers. Each type of dimmer has its advantages and limitations, depending on the specific application and lighting system. Some LED dimmers are designed for specific types of LED lighting systems, such as low-voltage or high-voltage LED lights, while others can work with a wide range of LED lights.

One important consideration when selecting an LED dimmer is compatibility with the LED light source. Not all LED lights are compatible with all types of dimmers, and using an incompatible dimmer can cause flickering, buzzing, or other performance issues. To ensure compatibility, it is important to select an LED dimmer that is specifically designed for the type of LED light being used.

LED dimmers are an important tool for interior designers, allowing them to create customized lighting schemes that enhance the look and feel of a space. By selecting the right type of dimmer and LED lighting system, designers can create dynamic and versatile lighting designs that meet the needs and preferences of their clients [11].

References

1. Enwin, A. D., Ikiriko, T. D., & Jonathan-Ihua, G. O. (2023). The role of colours in interior design of liveable spaces. *European Journal of Theoretical and Applied Sciences, 1*(4), 242–262.
2. Wardono, P., Hibino, H., & Koyama, S. (2012). Effects of interior colors, lighting and decors on perceived sociability, emotion and behavior related to social dining. *Procedia-Social and Behavioral Sciences, 38*, 362–372.

3. Mittal, N. (2021). *The key of interior design (Illustration of methods and principles)*. Standard Book House Since 1960.
4. Ching, F. D., & Binggeli, C. (2018). *Interior design illustrated*. John Wiley & Sons.
5. Mitton, M., & Nystuen, C. (2021). *Residential interior design: A guide to planning spaces*. John Wiley & Sons.
6. Karlen, M., Spangler, C., & Benya, J. R. (2017). *Lighting design basics*. John Wiley & Sons.
7. Kim, S. Y. (2005). LED effect: Interior element design. In *Proceedings of the Korean Institute of Interior Design conference* (pp. 275–278). Korean Institute of Interior Design.
8. Muhamad, W. N. W., Anuar, A. S. M., Sarnin, S. S., Azreen, M. M., Idris, A., & Kassim, M. (2018). Intelligence lighting control system with precise dimming and correlated color temperature. In *TENCON 2018–2018 IEEE region 10 conference* (pp. 1235–1240). IEEE.
9. Hua, J. (1999). Low voltage dimming system. In *Conference Record of the 1999 IEEE industry applications conference. Thirty-forth IAS annual meeting (Cat. No. 99CH36370)* (Vol. 3, pp. 1700–1704). IEEE.
10. Hui, S. R., Lee, L. M., Chung, H. H., & Ho, Y. K. (2001). An electronic ballast with wide dimming range, high PF, and low EMI. *IEEE Transactions on Power Electronics, 16*(4), 465–472.
11. Narra, P., & Zinger, D. S. (2004). An effective LED dimming approach. In *Conference record of the 2004 IEEE industry applications conference, 2004. 39th IAS annual meeting* (Vol. 3, pp. 1671–1676). IEEE.

Chapter 7
Professional Lighting for Interior Design

Contents

Professional lighting for interior design involves creating a well-designed and functional lighting scheme that meets the needs of the space and its occupants. It requires a thorough understanding of lighting principles, the latest lighting technologies and fixtures, and an ability to work with other design elements in the space. Professional lighting design can enhance the aesthetic appeal of the space, improve functionality, and promote energy efficiency.

Professional lighting designers work closely with architects, interior designers, and other design professionals to create lighting schemes that are tailored to the specific needs of the space. They typically consider factors such as the purpose of the space, the intended use of the space, the color scheme, and the type of furnishings and decor in the space. They also take into account the desired mood and atmosphere of the space and any specific requirements of the occupants, such as visual acuity needs or disability accommodations.

In addition to considering these factors, professional lighting designers use a variety of lighting technologies and fixtures to achieve the desired effect. This may include the use of LED lights, halogen lights, fluorescent lights, track lighting, recessed lighting, and other lighting sources. They may also use various lighting controls and systems to achieve the desired level of brightness, color temperature, and distribution of light.

G. Ozenen, *Architectural Interior Lighting*,
https://doi.org/10.1007/978-3-031-49695-0_7

Professional lighting designers are trained to balance the various elements of a lighting scheme, including brightness, color, contrast, and distribution of light. They understand the importance of proper lighting levels, color rendering, and glare control to promote visual comfort and health.

Overall, professional lighting design is an essential aspect of interior design that can greatly enhance the aesthetics, functionality, and energy efficiency of a space. By working with a professional lighting designer, designers can ensure that their lighting scheme meets the needs of the space and its occupants while creating an aesthetically pleasing and functional environment.

Professional lighting features in interior design can enhance the overall aesthetic and functionality of a space. Some of the key features of professional lighting for interior design include high-quality lighting fixtures, layered lighting, lighting controls, color temperature control, and lighting calculations.

- High-quality lighting fixtures: Professional lighting designers often use high-quality, designer fixtures that provide superior performance and visual appeal. These fixtures come in a variety of styles, sizes, and finishes to suit different design preferences.
- Layered lighting: Professional lighting designers create a layered lighting plan that combines different types of lighting, including ambient, task, and accent lighting. This approach helps create a balanced and functional lighting scheme that enhances the mood and functionality of a space.
- Lighting controls: Professional lighting designers often use lighting controls such as dimmers, timers, and occupancy sensors to improve energy efficiency and create customizable lighting options. These controls can be integrated with a building management system or used as standalone systems.
- Color temperature control: Color temperature is an important aspect of lighting design, and professional designers pay close attention to this aspect to ensure that the lighting complements the space and creates the desired atmosphere. Color temperature can be adjusted using different types of light sources, such as incandescent, fluorescent, and LED.
- Lighting calculations: Professional lighting designers use lighting calculations to determine the appropriate lighting levels for a space. These calculations take into account factors such as room size, ceiling height, and the type of activities that will be taking place in the space [1, 2].

7.1 High-Quality Lighting Fixtures

High-quality lighting fixtures are essential components of professional lighting design for interior spaces. These fixtures not only provide the necessary illumination for a space but also contribute to the aesthetic appeal and ambiance of the room. Here are some types of high-quality lighting fixtures commonly used in professional lighting design:

- Recessed Lighting: Recessed lighting fixtures are installed into the ceiling or wall, creating a seamless and clean look. They come in various sizes, shapes, and colors to match any interior design style.
- Pendant Lighting: Pendant lighting fixtures hang from the ceiling on a chain or cord and come in various sizes, shapes, and styles. They can be used as task lighting, accent lighting, or decorative lighting depending on the design of the fixture.
- Chandeliers: Chandeliers are ornate light fixtures that hang from the ceiling and feature multiple bulbs. They are typically used in grand entryways, dining rooms, and other spaces where a statement piece is desired.
- Wall Sconces: Wall sconces are fixtures that are mounted on the wall and provide an indirect, diffused light. They are often used in hallways, living rooms, and bedrooms as decorative or accent lighting.
- Table and Floor Lamps: Table and floor lamps are portable lighting fixtures that can be moved around a room to provide task or ambient lighting. They come in a variety of styles, sizes, and colors and can add a decorative element to a room.
- Track Lighting: Track lighting fixtures are mounted on a track that can be adjusted to direct light to different areas of a room. They are commonly used in kitchens, galleries, and retail spaces to highlight specific areas.

High-quality lighting fixtures offer several benefits, including energy efficiency, durability, and long-lasting performance. They also provide a range of design options, allowing designers to create unique and customized lighting schemes for any interior space [3].

7.2 Layered Lighting

Layered lighting is a technique used by lighting designers to create a balanced and aesthetically pleasing environment in interior design. It involves combining different types of lighting fixtures to provide multiple layers of light, each serving a specific purpose. This approach allows for greater flexibility in controlling the intensity, color, and direction of light to create different moods and enhance the visual interest of the space.

The three main layers of lighting in interior design are ambient, task, and accent lighting. Ambient lighting provides the general illumination of the space and creates a comfortable and functional environment. Task lighting is used to provide focused illumination for specific activities such as reading or cooking. Accent lighting is used to highlight and enhance the visual interest of specific features, such as artwork or architectural elements.

In professional lighting design, layered lighting is achieved by combining different types of light fixtures, such as chandeliers, pendant lights, wall sconces, table and floor lamps, and LED strip lights, among others. The fixtures are strategically placed in the space to achieve the desired effect and create a cohesive and visually appealing design.

High-quality lighting fixtures are essential for achieving optimal results in professional lighting design. Quality fixtures provide superior light output, color rendering, and energy efficiency, and have a longer lifespan. The choice of fixtures depends on the specific requirements of the project, such as the size and function of the space, the desired ambiance, and the budget.

7.3 Lighting Controls

Lighting controls are an essential aspect of professional lighting design as they allow for flexible and dynamic lighting schemes. The use of lighting controls enables designers to create multiple lighting scenes, adjust light levels, and incorporate daylight harvesting strategies to reduce energy consumption. Lighting controls can be divided into two main categories: manual and automated.

Manual lighting controls include dimmers, switches, and timers. Dimmers allow users to adjust light levels, while switches turn lights on and off. Timers can be used to automate lighting schedules and turn lights on and off at specific times. Manual controls are often used in residential settings and smaller commercial spaces.

Automated lighting controls, on the other hand, use sensors and programming to adjust lighting levels based on occupancy, daylight levels, and time of day. Examples of automated controls include occupancy sensors, daylight sensors, and timeclocks. Occupancy sensors detect when a space is occupied and turn lights on and off accordingly. Daylight sensors adjust artificial light levels based on the amount of natural light available in a space. Timeclocks can be programmed to turn lights on and off at specific times, providing additional energy savings.

In addition to these basic control types, more advanced systems can be used to create complex lighting scenes and effects. For example, DMX control systems can be used to create dynamic lighting displays for entertainment and hospitality venues. These systems allow for precise control over individual light fixtures and can be programmed to create complex lighting scenes and effects.

Overall, the use of lighting controls is a critical component of professional lighting design, allowing designers to create dynamic and energy-efficient lighting schemes that enhance the function and aesthetics of interior spaces [4].

7.4 Color Temperature Control

Color temperature control is a crucial aspect of professional lighting features in interior design. It refers to the ability to adjust the warmth or coolness of the light emitted by the light source. The color temperature is measured in Kelvin (K), and it affects the ambiance, mood, and perception of a space.

In interior design, color temperature control is used to create different atmospheres and moods in different areas of a space. For instance, warmer color temperatures are ideal for creating a cozy and intimate feel, while cooler temperatures can be used to create a bright and energetic atmosphere.

One way to achieve color temperature control is through the use of LED lighting. LED lights come in different color temperatures, ranging from warm white (around 2700 K) to cool white (around 6500 K), and can be easily controlled using a dimmer switch or a smart lighting system.

Another way to achieve color temperature control is through the use of color-changing LED lights, which can be programmed to change color temperature over time or with the touch of a button. This allows for greater flexibility and creativity in lighting design.

In addition, color temperature control can also be achieved through the use of color filters or gels on traditional light sources, such as incandescent or fluorescent lights. These filters can change the color temperature of the light emitted by the source.

Overall, color temperature control is an essential aspect of professional lighting features in interior design, allowing designers to create customized lighting solutions that enhance the mood, ambiance, and functionality of a space [4].

7.5 Lighting Calculations

Lighting calculations are an essential aspect of professional lighting design, allowing designers to determine the appropriate light levels, fixture placement, and wattage required to achieve the desired effect. There are several calculations involved in lighting design, including the following:

- Lumen Method: The lumen method is used to determine the number of luminaires required to achieve a specific light level in a space. It involves calculating the total lumens required based on the room size, type of space, and activity level and then dividing by the lumens per fixture to determine the number of fixtures needed.
- Point-by-Point Method: The point-by-point method involves calculating the illumination levels at specific points in the space based on the height of the ceiling, the distance between the fixtures, and the type of fixture being used.
- Zonal Cavity Method: The zonal cavity method involves calculating the light levels in the space by dividing it into zones and determining the lumens required for each zone based on the activity level and the reflectance of the surfaces in the zone.
- Daylighting Calculations: Daylighting calculations are used to determine the amount of natural light entering a space and how it can be used to supplement artificial lighting. These calculations take into account the orientation of the building, the size and location of windows, and the amount of sunlight in the area.

Lighting calculations can be complex and require specialized knowledge, software, and equipment. Working with a professional lighting designer can ensure that the calculations are accurate and result in the most efficient and effective lighting design for the space [4, 5].

7.5.1 Lumen Method

The lumen method is a commonly used lighting calculation method in professional lighting design. It is used to determine the number of light fixtures needed to achieve a desired level of illumination in a given space. The method takes into account the size of the space, the height of the ceiling, and the reflectance of the surfaces in the room.

The lumen method involves calculating the total lumen output required for a space based on its size and the desired level of illumination. This calculation takes into account the reflectance of the surfaces in the room, as well as any daylight entering the space. Once the required lumen output is determined, it is divided by the lumen output of the light fixtures being used to determine the number of fixtures needed.

There are different variations of the lumen method, such as the point-by-point method and the zonal cavity method, which take into account the distribution of light in the space and the angles of incidence of light on surfaces.

Lighting designers may use computer software programs to perform lighting calculations and determine the optimal lighting design for a space. These programs can take into account a variety of factors, such as daylighting, energy efficiency, and the color rendering index (CRI) of different light sources [6].

7.5.2 Point-by-Point Method

The point-by-point method is a lighting calculation technique used in professional lighting design to ensure that a space is properly illuminated. This method involves calculating the light output required for each individual point in the space, taking into account the position of the light fixtures, the reflectance of surfaces, and the desired light levels.

To use the point-by-point method, designers must first establish the illuminance level required for the space, based on the activities that will take place in the area. They must then identify the types of fixtures that will be used, the mounting heights and locations, and the beam angles of each fixture. The next step is to determine the lumens output required for each fixture, based on the illuminance level and the area to be illuminated.

Once the lumens output has been established, designers must take into account the reflectance of the surfaces in the space, as this affects the amount of light that is actually delivered to the task surface. Finally, they can calculate the number of fixtures required to achieve the desired illuminance level at each point in the space.

The point-by-point method is a more precise technique than the lumen method, which involves calculating the total number of lumens required to illuminate a space, based on the square footage of the area and the desired illuminance level. The point-by-point method takes into account the specific requirements of each area of the space, resulting in a more accurate and efficient lighting design [7].

7.5.3 *Zonal Cavity Method*

The zonal cavity method is a lighting calculation method used to determine the illuminance at any point in a space by considering the interaction of all luminaires and surfaces. It is a widely accepted and recognized method for lighting design in architectural spaces.

The zonal cavity method divides a space into a series of zones based on the geometry of the luminaires and the surfaces they illuminate. Each zone is characterized by its reflectance and position relative to the luminaires. By calculating the amount of light that each luminaire contributes to each zone, the total illuminance at any point in the space can be determined.

The zonal cavity method takes into account the reflective properties of all surfaces in the space, including walls, ceilings, floors, and any other surfaces that may affect the lighting. This method considers the absorption and reflection of light by the surfaces, as well as the shape and size of the space, and the distribution of the luminaires.

The zonal cavity method can be used for a variety of lighting design applications, including calculating illuminance levels for general lighting, task lighting, and accent lighting. It is also useful for determining the optimal placement and configuration of luminaires to achieve a desired level of illumination and aesthetic effect.

In summary, the zonal cavity method is a comprehensive lighting calculation method that takes into account the geometry of the space, the reflective properties of surfaces, and the distribution of luminaires to determine illuminance levels throughout a space [8].

7.5.4 *Daylighting Calculations*

Daylighting calculations are an essential part of lighting design and help determine how much natural light can be used to illuminate a space. By taking into account the location, orientation, and size of windows, as well as external factors such as the position of the sun, daylighting calculations can be used to optimize the use of natural light in a building, reduce energy consumption, and create a more comfortable and productive environment for occupants.

One common method for daylighting calculations is the daylight factor method, which calculates the ratio of light available inside a room to the light available outside. This can be used to determine the optimum size and location of windows, as well as the placement of interior surfaces such as walls and ceilings to reflect and distribute natural light.

Other methods for daylighting calculations include the use of computer simulations and physical models, which can provide more detailed and accurate results but require more time and resources. Additionally, daylighting calculations may be required by building codes or certification programs such as LEED (Leadership in Energy and Environmental Design) to ensure that buildings meet certain energy efficiency and sustainability standards [3, 4].

References

1. Livingston, J. (2021). *Designing with light: The art, science, and practice of architectural lighting design*. Wiley.
2. Michel, L. (1995). *Light: The shape of space: Designing with space and light*. Wiley.
3. Skowranek, R. (2017). *Basics lighting design*. Birkhäuser.
4. Karlen, M., Spangler, C., & Benya, J. R. (2017). *Lighting design basics*. Wiley.
5. Tregenza, P., & Loe, D. (2013). *The design of lighting*. Routledge.
6. Biesele, R. L., Arner, W. J., & Conover, E. W. (1953). A lumen method of daylighting design. *Illuminating Engineering Journal, 48*(1), 39–45.
7. Martinez, A., Khrushchev, I. Y., & Bennion, I. (2005). Thermal properties of fibre Bragg gratings inscribed point-by-point by infrared femtosecond laser. *Electronics Letters, 41*(4), 176–178.
8. Parent, M. D., & Murdoch, J. B. (1988). The expansion of the zonal cavity method of interior lighting design to include skylights. *Journal of the Illuminating Engineering Society, 17*(2), 141–173.

Chapter 8
Lighting Controls and Systems

Contents

Lighting controls and systems refer to various devices, technologies, and strategies used to manage the output and performance of lighting systems. These controls can be used to regulate the intensity, color, and distribution of light to improve energy efficiency, enhance user comfort, and reduce operating costs.

Some examples of lighting controls and systems are as follows:

- Dimming systems that can adjust the light output of luminaires based on user needs or time of day
- Motion sensors that detect occupancy and turn lights on or off automatically

G. Ozenen, *Architectural Interior Lighting*,
https://doi.org/10.1007/978-3-031-49695-0_8

- Daylight sensors that adjust light output based on the amount of natural light available in a space
- Timeclocks that turn lights on or off according to a pre-determined schedule
- Networked lighting controls that allow for centralized management and monitoring of lighting systems
- Wireless controls that enable flexible installation and programming of lighting systems.

Lighting controls and systems are an important component of architectural interior lighting design, as they offer a range of benefits to both building owners and occupants, including energy savings, improved comfort and productivity, and greater flexibility in lighting design [1, 2].

Types of Lighting controls and systems are as follows:

- On/off switches
- Dimmer switches
- Occupancy sensors
- Daylight sensors
- Timers
- Lighting control panels
- Building automation systems (BAS)
- Centralized lighting control systems
- Wireless lighting control systems
- Internet of Things (IoT)-enabled lighting control systems
- Software-based lighting control systems
- Emergency lighting control systems
- Energy management systems (EMS) for lighting
- Light-emitting diode (LED) drivers and controllers
- Smart lighting systems with voice and/or motion control

8.1 On/Off Switches

On/off switches are one of the most basic types of lighting controls, allowing users to turn lights on and off as needed. These switches can be used to control individual lights, groups of lights, or entire lighting systems. On/off switches are commonly found in residential, commercial, and industrial settings and are available in a variety of styles and designs to suit different applications and aesthetics.

One study found that the use of on/off switches in lighting control systems can significantly reduce energy consumption by allowing users to turn off lights when they are not needed, such as during daylight hours or when a space is unoccupied. Another study found that the use of on/off switches in combination with occupancy sensors can further enhance energy savings by automatically turning off lights when a space is unoccupied.

In addition to their energy-saving benefits, on/off switches can also improve the comfort and convenience of lighting systems by giving users greater control over the lighting environment. For example, they can be used to create different lighting scenes for different activities or times of day, such as dimming lights in the evening for a more relaxing atmosphere.

Overall, on/off switches are a simple yet effective lighting control option that can offer significant energy savings and improved user experience when integrated into lighting systems [3].

8.2 Dimmer Switches

Dimmer switches are devices that allow the user to adjust the level of light output from a light source. They are commonly used in residential and commercial lighting applications to create ambiance and save energy by reducing the amount of light needed in a given space.

There are several types of dimmer switches available, including rheostat, autotransformer, magnetic, and electronic dimmers. Rheostat and autotransformer dimmers are older technologies that are less commonly used today, while magnetic and electronic dimmers are more prevalent.

Magnetic dimmers use a transformer to vary the amount of voltage supplied to the light source, while electronic dimmers use a solid-state device to control the amount of power delivered to the light. Electronic dimmers are more efficient and offer better control over the range of light output.

Some of the benefits of using dimmer switches include energy savings, extended bulb life, and increased comfort and convenience. Dimmer switches can be used with a variety of light sources, including incandescent, halogen, fluorescent, and LED bulbs [3].

8.3 Occupancy Sensors

Occupancy sensors are devices that detect the presence of people in a room and automatically turn the lights on or off based on their occupancy status. They are commonly used in spaces such as offices, conference rooms, restrooms, and storage areas to save energy by automatically turning off the lights when the space is unoccupied.

There are two types of occupancy sensors: ultrasonic and passive infrared (PIR). Ultrasonic sensors emit high-frequency sound waves that bounce off objects in the room, while PIR sensors detect the infrared energy emitted by a person's body heat.

Occupancy sensors can be standalone devices that plug into an outlet or wired into a lighting control system. They can be set to turn the lights off after a certain amount of time has elapsed since the last detected occupancy, known as the "time-out" period.

In addition to saving energy, occupancy sensors can also improve the user experience by providing automatic lighting control [4].

8.4 Daylight Sensors

Daylight sensors are devices that automatically adjust artificial lighting levels based on the amount of natural daylight available in a space. These sensors can be used to reduce energy consumption and promote the use of natural light in buildings.

Daylight sensors typically consist of a photosensor, which measures the amount of daylight present, and a control device, which adjusts the lighting system accordingly. Some daylight sensors can also be programmed to adjust the lighting levels based on the time of day, season, or weather conditions.

Studies have shown that the use of daylight sensors in lighting systems can lead to significant energy savings [5].

8.5 Timers

Timers are another type of lighting control that are used to turn lights on and off at pre-determined times. Timers can be set to turn lights on and off at specific times of day or night, or they can be set to turn lights on and off after a certain period of time has elapsed.

Timers are commonly used in outdoor lighting systems to control the operation of security lights, landscape lighting, and other types of outdoor lighting fixtures. They are also used in indoor lighting systems to control the operation of lights in areas that are not frequently used, such as storerooms, closets, and other utility areas.

There are different types of timers available for lighting control, including mechanical timers, digital timers, and smart timers. Mechanical timers are the most basic type of timer and use a simple mechanical mechanism to turn lights on and off at pre-determined times. Digital timers, on the other hand, use an electronic timer to control the operation of lights and can be programmed with more precision than mechanical timers. Smart timers are the most advanced type of timer and use Wi-Fi or other wireless connectivity to allow users to control their lighting systems from their smartphones or other mobile devices [6].

8.6 Lighting Control Panels

Lighting control panels are centralized systems used for managing and controlling lighting in a building. They are designed to control multiple circuits of lighting simultaneously and provide a range of control options to ensure that lighting levels are optimized for different scenarios, such as time of day, occupancy, and activity. Lighting control panels can be customized to meet the specific needs of different buildings and can be programmed to operate in various modes, such as manual, automatic, or semi-automatic.

Lighting control panels are typically installed in a dedicated room or closet, and can be integrated with other building systems, to create a comprehensive building automation system. Some lighting control panels use specialized software to enable remote monitoring and control via a web-based interface, allowing facility managers to monitor and adjust lighting levels in real time [7].

8.7 Building Automation Systems (BAS)

Building automation systems (BAS) are computerized control systems that manage various building systems, including lighting, heating, ventilation, and air conditioning. BAS can be used to create a fully integrated and automated building, allowing for greater energy efficiency, increased comfort, and improved building performance. Lighting control is one of the key functions of a BAS.

A BAS typically includes a central computer system that communicates with various building systems, including lighting controls. The central computer system can be programmed to automatically adjust lighting levels based on occupancy, time of day, and other factors. In addition, the system can be used to monitor and analyze energy usage and to identify opportunities for further energy savings.

BAS can be customized to meet the specific needs of a building and its occupants. For example, a BAS can be programmed to adjust lighting levels based on occupancy in different parts of a building or to provide individual control of lighting for occupants in certain areas [8–10].

8.8 Centralized Lighting Control Systems

Centralized lighting control systems are a type of lighting control system that allows for the centralized control of multiple lighting zones or circuits from a central location. These systems typically use a central control panel or software to communicate with individual lighting controllers, allowing for greater flexibility and control over lighting settings.

One of the primary benefits of centralized lighting control systems is their ability to manage and optimize energy usage. By monitoring and adjusting lighting settings based on occupancy, time of day, and other factors, these systems can significantly reduce energy consumption and associated costs.

Examples of centralized lighting control systems include DALI (Digital Addressable Lighting Interface) and DMX (Digital Multiplex) systems. DALI is a standardized protocol for lighting control that allows for two-way communication between the central control panel and individual lighting fixtures or zones. DMX, on the other hand, is a protocol commonly used in entertainment and stage lighting that can also be adapted for architectural lighting control [11].

8.9 Wireless Lighting Control Systems

Wireless lighting control systems are a type of lighting control system that allows for remote control of lighting fixtures without the need for wired connections. These systems typically use radio frequency (RF) or infrared (IR) signals to communicate between control devices and lighting fixtures.

One advantage of wireless lighting control systems is their flexibility in installation and reconfiguration. Since there are no wires to run, these systems can be installed more quickly and easily, and they can be easily reconfigured if lighting needs change.

Another advantage is that wireless systems can be integrated with other building automation systems, such as HVAC or security systems, to provide centralized control of all building systems.

Some examples of wireless lighting control systems include Philips Hue, Lutron Caseta, and Leviton Decora Smart. These systems use different technologies and protocols for communication, such as Zigbee, Bluetooth, or Wi-Fi [12, 13].

8.10 Internet of Things (IoT)-Enabled Lighting Control Systems

Internet of Things (IoT)-enabled lighting control systems are becoming increasingly popular due to their ability to integrate lighting control with other building systems and allow for remote access and control. These systems use sensors and other devices to collect data about the environment and adjust lighting accordingly, making them highly energy-efficient.

One of the main benefits of IoT-enabled lighting control systems is their ability to provide granular control over individual fixtures or groups of fixtures. This allows for more precise adjustments based on occupancy levels, time of day, and other factors. In addition, these systems can be programmed to automatically adjust lighting levels based on changes in natural light, reducing energy consumption and improving user comfort.

IoT-enabled lighting control systems also offer the potential for advanced analytics and data-driven insights into energy usage and occupancy patterns. By collecting data from sensors and other devices, these systems can provide valuable information for building owners and operators to optimize energy efficiency and improve the overall user experience.

However, there are also concerns around the security and privacy implications of IoT-enabled lighting control systems, as they rely on a network of connected devices that can potentially be vulnerable to cyberattacks [14, 15].

8.11 Software-Based Lighting Control Systems

Software-based lighting control systems utilize computer programs to manage lighting settings and provide more advanced features such as scheduling, zoning, and dimming controls. These systems allow for remote control and monitoring of lighting, providing convenience and energy efficiency.

One common software-based lighting control system is Digital Addressable Lighting Interface (DALI), which is a protocol used to communicate with and control dimming ballasts and LED drivers in a lighting system. DALI allows for individual control of lighting fixtures and offers the ability to set different lighting scenes and schedules.

Another example of a software-based lighting control system is Lighting Control and Design's LIGHTSYNC software, which offers a user-friendly interface for programming and controlling lighting systems. The software allows for real-time monitoring and adjustment of lighting settings and integrates with other building automation systems for centralized control [16].

8.12 Emergency Lighting Control Systems

Emergency lighting control systems are designed to provide illumination in the event of a power outage or emergency situation. They are typically installed in public buildings, such as schools, hospitals, and commercial offices, where the safety of occupants is of paramount concern. These systems are often required to meet regulatory standards and codes to ensure the safety of the occupants.

There are several types of emergency lighting control systems available, including the following:

- Self-contained systems: These systems are designed to operate independently of the building's power supply. They incorporate a battery backup and charger to provide power to the emergency lights during a power outage.

- Centralized systems: These systems are connected to a centralized power source and are designed to provide power to multiple emergency lights in a building. They are typically used in larger buildings, such as hospitals or shopping centers.
- Self-contained and centralized combination systems: These systems combine the features of self-contained and centralized systems. They are designed to provide power to a limited number of emergency lights, typically located in critical areas of the building.

Emergency lighting control systems can be further categorized based on the type of emergency lighting they control. This includes the following:

- Exit signs: These systems control the illumination of exit signs in a building to ensure that they are visible in the event of an emergency.
- Emergency lights: These systems control the illumination of emergency lights in a building, including those installed in stairwells, corridors, and other critical areas.
- Exit signs and emergency lights combination systems: These systems control both exit signs and emergency lights [17].

8.12.1 Self-Contained Emergency Lighting Control Systems

Self-contained lighting control systems are those that incorporate all the necessary control components within a single device or module. These systems are typically designed for specific applications, such as occupancy sensing or daylight harvesting, and offer a simplified approach to lighting control. Self-contained systems are often used in retrofit projects, where it may be impractical to install a more complex control system.

One common type of self-contained system is the occupancy sensor. These devices use infrared or ultrasonic technology to detect the presence of people in a space and can turn lights on or off accordingly. They are particularly useful in spaces that are not frequently occupied, such as storage rooms or restrooms, where lights may be left on unnecessarily.

Another type of self-contained system is the daylight sensor. These devices use photocells to measure the amount of natural light in a space and can adjust artificial lighting levels accordingly. This can result in significant energy savings, particularly in spaces with large windows or skylights.

Some self-contained systems also incorporate wireless communication, allowing for easy installation and integration with other building systems. For example, some occupancy sensors can communicate with a central building automation system, allowing for more centralized control and monitoring [18].

8.12.2 Centralized Emergency Lighting Control Systems

Centralized emergency lighting control systems are used to monitor and control emergency lighting systems in large buildings or complexes from a central location. They provide a central point of control for emergency lighting systems, which can be critical in emergency situations. These systems are typically used in buildings such as hospitals, hotels, office buildings, and other large public spaces where a loss of power could cause serious harm to occupants.

Centralized emergency lighting control systems can be configured to automatically turn on emergency lights in the event of a power outage or other emergency situation. They can also be programmed to perform routine testing and maintenance of the emergency lighting system to ensure that it is functioning properly. Some systems can even alert maintenance personnel when a light bulb needs to be replaced or when a battery needs to be replaced.

Centralized emergency lighting control systems typically consist of a control panel, which is connected to the emergency lighting fixtures throughout the building. The control panel can be programmed to activate specific lights or zones in the event of an emergency and can also be used to test and maintain the emergency lighting system [19].

8.12.3 Self-Contained and Centralized Combination Systems

Self-contained and centralized combination systems in lighting control and systems provide a comprehensive solution to emergency lighting needs. These systems use a combination of self-contained emergency lights and centralized control systems to provide lighting in the event of a power failure or emergency situation.

Self-contained emergency lights are standalone devices that have a battery backup and provide emergency lighting without relying on any external power source. These lights are typically used in smaller spaces, such as individual rooms or hallways.

Centralized control systems, on the other hand, are designed to control multiple emergency lights from a central location. These systems typically use a central control panel to monitor the status of emergency lights, as well as to perform routine testing and maintenance.

Combining these two systems can provide a more comprehensive emergency lighting solution that is both reliable and efficient. The self-contained emergency lights provide localized lighting in the event of a power failure, while the centralized control system ensures that all lights are functioning properly and can be easily tested and maintained.

One example of a self-contained and centralized combination system is the Lithonia Lighting® EU2L Emergency LED Light. This device features a self-contained battery backup system that provides emergency lighting in the event of a power failure, as well as a centralized control system that allows for remote testing and maintenance [20].

8.12.4 Exit Signs

Exit signs are an important component of emergency lighting systems in buildings. These signs are designed to provide clear and visible guidance for occupants to exit a building safely during an emergency, such as a fire or power outage. Exit signs are typically illuminated to ensure their visibility even in low light conditions.

There are different types of exit signs available, including traditional incandescent bulb signs, LED signs, and photoluminescent signs. LED exit signs are becoming increasingly popular due to their energy efficiency, longer lifespan, and brighter illumination. Photoluminescent exit signs are also gaining popularity as they do not require electricity and can continue to glow even in complete darkness for several hours.

In addition to the type of sign, the placement and installation of exit signs are also critical for their effectiveness [21].

8.12.5 Emergency Lights

Emergency lights are an essential component of emergency lighting control systems. They provide illumination during power outages or other emergencies, helping occupants of a building to navigate safely to exits. Emergency lights can be installed as standalone fixtures or as part of a larger emergency lighting system.

There are several types of emergency lights available on the market, including the following:

1. Battery-operated emergency lights: These lights are self-contained units that are powered by rechargeable batteries. They can be mounted on walls or ceilings and are typically used in small commercial or residential buildings.
2. Emergency lights with remote heads: These lights have a central power source that is connected to remote light heads via low-voltage wiring. This allows the lights to be spread out over a larger area, providing more comprehensive coverage.
3. Central battery system (CBS) emergency lights: CBS systems are designed to power multiple emergency lights from a central location. They are typically used in larger commercial or industrial buildings.
4. Self-testing emergency lights: These lights are equipped with self-testing capabilities that allow them to automatically perform diagnostic tests to ensure they are functioning properly. This reduces the need for manual testing and maintenance.

It is important to ensure that emergency lights are installed and maintained in compliance with these regulations to ensure maximum safety for building occupants [21].

8.12.6 Exit Signs and Emergency Lights Combination Systems

Exit signs and emergency lights combination systems are an essential part of building safety systems, as they help occupants find their way out of a building in the event of an emergency. These systems are typically required by building codes and standards, such as the National Fire Protection Association (NFPA) Life Safety Code.

Exit signs are typically installed above doors and other exit routes to indicate the direction of the nearest exit. They are typically illuminated using LEDs or other energy-efficient light sources and may have battery backup in case of power failure. Emergency lights are designed to provide backup lighting in case of power failure, helping occupants see their way to safety.

Combination of exit sign and emergency light systems provides both functions in a single unit, reducing the overall number of devices needed to comply with safety codes. These systems can be self-contained, with a battery backup and control circuitry built into each unit, or centralized, with a separate control panel that manages multiple units throughout a building [22].

8.13 Energy Management Systems (EMS) for Lighting

Energy management systems (EMS) for lighting refer to systems that help manage and control lighting usage in a building, with the aim of optimizing energy efficiency and reducing energy consumption. EMS for lighting can be used in both residential and commercial settings and typically include features such as scheduling, occupancy sensors, and daylight sensors, among others.

One of the key components of EMS for lighting is the use of advanced lighting control systems that incorporate intelligent controls and sensors. These systems allow for precise control of lighting levels, based on occupancy patterns and daylight levels, among other factors. They also enable centralized control and monitoring of lighting systems, which can help identify areas where energy savings can be made.

Another important feature of EMS for lighting is the use of energy monitoring and reporting tools. These tools allow building owners and managers to track energy usage and identify areas where improvements can be made. For example, energy monitoring tools can help identify when lighting usage is highest and provide insights into how lighting usage patterns can be optimized to reduce energy consumption.

Overall, EMS for lighting can provide significant benefits for building owners and managers, including reduced energy consumption, improved occupant comfort and productivity, and lower energy costs. With the increasing focus on sustainability and energy efficiency, the adoption of EMS for lighting is likely to continue to grow in the coming years [23, 24].

8.14 Light-Emitting Diode (LED) Drivers and Controllers

LEDs are becoming increasingly popular as a light source due to their energy efficiency, long lifespan, and low maintenance requirements. However, in order to properly power and control LEDs, specialized LED drivers and controllers are needed.

LED drivers convert incoming AC voltage to the DC voltage required to power LEDs. They also regulate the current and voltage supplied to the LED, ensuring stable and consistent performance. LED drivers can be either constant voltage or constant current, depending on the type of LED being used and the requirements of the application.

LED controllers, on the other hand, allow for the dimming and color control of LED lights. They can be either wired or wireless and offer a wide range of functionality including the ability to control multiple zones or groups of lights, set schedules and timers, and adjust color temperature and brightness.

Some popular LED driver and controller brands in the lighting industry include Philips Hue, Lutron, Leviton, and Cree [25].

8.15 Smart Lighting Systems with Voice and/or Motion Control

Smart lighting systems with voice and/or motion control are becoming increasingly popular as they offer a convenient and energy-efficient way to control lighting in buildings. These systems use voice commands or motion sensors to turn lights on or off, adjust their brightness, or change their color.

Voice-controlled lighting systems use voice-activated virtual assistants, such as Amazon Alexa or Google Assistant, to control lights in a building. Users can simply speak voice commands, such as "turn on the lights" or "dim the lights," and the system will respond accordingly.

Motion-controlled lighting systems use sensors that detect movement to turn lights on or off. These systems can be programmed to turn lights on when someone enters a room and turn them off when the room is empty. This can help save energy by ensuring that lights are not left on when they are not needed.

Some smart lighting systems use both voice and motion control to provide a more comprehensive and flexible lighting control solution. These systems can be programmed to respond to specific voice commands and to activate or deactivate certain lights based on motion detection.

Smart lighting systems with voice and/or motion control can be integrated into other building automation systems, such as HVAC or security systems, to create a fully automated and efficient building environment [26].

References

1. DiLouie, C. (2021). *Advanced lighting controls: Energy savings, productivity, technology and applications*. CRC Press.
2. Benediktsson, G. (2009). *Lighting control: Possibilities in cost and energy-efficient lighting control techniques*. LUTEDX/TEIE.
3. Reinhart, C. F., & Jones, C. (2004). Electric lighting energy savings for a photocell controlled dimmer – A comparative simulation study using DOE.2 and Lightswitch Wizard. In *Proceedings of the ESIM 2004 conference, Vancouver, BC* (pp. 183–189). Institute for Research in Construction.
4. Delaney, D. T., O'Hare, G. M., & Ruzzelli, A. G. (2009). Evaluation of energy-efficiency in lighting systems using sensor networks. In *Proceedings of the first ACM workshop on embedded sensing systems for energy-efficiency in buildings* (pp. 61–66). ACM.
5. Li, S., Pandharipande, A., & Willems, F. M. (2016). Daylight sensing LED lighting system. *IEEE Sensors Journal, 16*(9), 3216–3223.
6. Chen, K., & Castenachiold, R. (1985). Selecting lighting controls for optimum energy savings. In *1985 Annual meeting industry applications society* (pp. 1697–1705). IEEE.
7. De Paz, J. F., Bajo, J., Rodríguez, S., Villarrubia, G., & Corchado, J. M. (2016). Intelligent system for lighting control in smart cities. *Information Sciences, 372*, 241–255.
8. Domingues, P., Carreira, P., Vieira, R., & Kastner, W. (2016). Building automation systems: Concepts and technology review. *Computer Standards & Interfaces, 45*, 1–12.
9. Kastner, W., Neugschwandtner, G., Soucek, S., & Newman, H. M. (2005). Communication systems for building automation and control. *Proceedings of the IEEE, 93*(6), 1178–1203.
10. Wong, A. C. W., & So, A. T. P. (1997). Building automation in the 21st century. In *1997 Fourth international conference on advances in power system control, operation and management, APSCOM-97*. IET.
11. Pandharipande, A., Rossi, M., Caicedo, D., Schenato, L., & Cenedese, A. (2015). Centralized lighting control with luminaire-based occupancy and light sensing. In *2015 IEEE 13th international conference on industrial informatics (INDIN)* (pp. 31–36). IEEE.
12. Pandharipande, A., Caicedo, D., & Wang, X. (2014). Sensor-driven wireless lighting control: System solutions and services for intelligent buildings. *IEEE Sensors Journal, 14*(12), 4207–4215.
13. Pandharipande, A., & Li, S. (2013). Light-harvesting wireless sensors for indoor lighting control. *IEEE Sensors Journal, 13*(12), 4599–4606.
14. Sikder, A. K., Acar, A., Aksu, H., Uluagac, A. S., Akkaya, K., & Conti, M. (2018). IoT-enabled smart lighting systems for smart cities. In *2018 IEEE 8th annual computing and communication workshop and conference (CCWC)* (pp. 639–645). IEEE.
15. Hossein Motlagh, N., Mohammadrezaei, M., Hunt, J., & Zakeri, B. (2020). Internet of Things (IoT) and the energy sector. *Energies, 13*(2), 494.

16. Ordaz-García, O. O., Ortiz-López, M., Quiles-Latorre, F. J., Arceo-Olague, J. G., Solís-Robles, R., & Bellido-Outeiriño, F. J. (2020). DALI bridge FPGA-based implementation in a wireless sensor node for IoT street lighting applications. *Electronics, 9*(11), 1803.
17. Tse, W. L., Chan, W. L., & Lai, S. S. (2003). Emergency lighting monitoring system using LonWorks. *Automation in Construction, 12*(5), 617–629.
18. Tiana, C. (2007). Compact self-contained enhanced-vision system (EVS) sensor simulator. In *Enhanced and synthetic vision 2007* (Vol. 6559, pp. 201–207). SPIE.
19. Barker, H. P. (1938). The centralized control of public lighting and off-peak loads by superimposed ripples. *Journal of the Institution of Electrical Engineers, 83*(504), 823–836.
20. DiLouie, C. (2020). *Lighting controls handbook*. CRC Press.
21. Wong, L. T., & Lo, K. C. (2007). Experimental study on visibility of exit signs in buildings. *Building and Environment, 42*(4), 1836–1842.
22. Ouellette, M. J., & Rea, M. S. (1989). Illuminance requirements for emergency lighting. *Journal of the Illuminating Engineering Society, 18*(1), 37–42.
23. Lee, D., & Cheng, C. C. (2016). Energy savings by energy management systems: A review. *Renewable and Sustainable Energy Reviews, 56*, 760–777.
24. Al-Ghaili, A. M., Kasim, H., Al-Hada, N. M., Jørgensen, B. N., Othman, M., & Wang, J. (2021). Energy management systems and strategies in buildings sector: A scoping review. *IEEE Access, 9*, 63790–63813.
25. Yeh, N., & Chung, J. P. (2009). High-brightness LEDs—Energy efficient lighting sources and their potential in indoor plant cultivation. *Renewable and Sustainable Energy Reviews, 13*(8), 2175–2180.
26. Prange, S., Von Zezschwitz, E., & Alt, F. (2019). Vision: Exploring challenges and opportunities for usable authentication in the smart home. In *2019 IEEE European symposium on security and privacy workshops (EuroS&PW)* (pp. 154–158). IEEE.

Chapter 9
Sustainable Lighting Design

Contents

Sustainable lighting design is an approach to designing lighting systems that prioritizes energy efficiency, environmental responsibility, and human well-being. It involves using techniques and technologies that reduce energy consumption and

G. Ozenen, *Architectural Interior Lighting*,
https://doi.org/10.1007/978-3-031-49695-0_9

minimize negative environmental impacts, while still providing adequate lighting for people to perform their tasks comfortably and safely.

One key aspect of sustainable lighting design is using energy-efficient light sources such as light-emitting diodes (LEDs) and compact fluorescent lamps (CFLs) instead of traditional incandescent bulbs. These sources use less energy and have longer lifetimes, reducing waste and the need for frequent replacements. In addition, sustainable lighting design often incorporates daylighting strategies, such as using natural light to supplement or replace electric lighting, as well as lighting controls that allow users to adjust lighting levels based on their needs [1, 2].

Other elements of sustainable lighting design can include using recycled or renewable materials in fixtures and systems, minimizing light pollution by controlling the direction and intensity of light, and incorporating lighting into a building's overall energy management strategy.

There are some key features of sustainable lighting design:

- Energy efficiency: Sustainable lighting design prioritizes energy efficiency to reduce energy consumption and minimize the environmental impact of lighting systems.
- Use of renewable energy: The use of renewable energy sources such as solar, wind, and geothermal power is encouraged to power lighting systems.
- Daylight harvesting: Sustainable lighting design integrates daylighting strategies to make use of natural light, reducing the need for artificial lighting.
- Light pollution reduction: Sustainable lighting design reduces light pollution by minimizing the amount of light that spills into the sky or into neighboring properties.
- Use of environmentally friendly materials: Sustainable lighting design incorporates the use of environmentally friendly materials such as recycled and biodegradable materials.
- Long lifespan: Sustainable lighting design emphasizes the use of lighting systems with a long lifespan, reducing waste and the need for frequent replacements.
- Lighting controls: Sustainable lighting design uses lighting controls to adjust lighting levels based on occupancy, daylight availability, and user preferences, reducing energy consumption.
- Human-centric lighting: Sustainable lighting design takes into account the health and well-being of occupants by using lighting that supports circadian rhythms and promotes productivity.
- Maintenance and serviceability: Sustainable lighting design prioritizes systems that are easy to maintain and service, reducing waste and promoting sustainability [1–3].

9.1 Energy Efficiency

Energy efficiency is an important aspect of sustainable lighting design. It involves designing lighting systems that consume less energy while still meeting the lighting needs of the space. There are several ways to achieve energy efficiency in lighting design, including the use of efficient light sources, proper lighting controls, and designing for daylighting.

One of the most effective ways to improve energy efficiency in lighting design is by using efficient light sources. Light-emitting diodes (LEDs) are currently the most efficient light source available, with energy savings of up to 80% compared to traditional incandescent bulbs. LED technology is constantly improving, with higher efficiencies and better color rendering becoming available. Additionally, LEDs have a longer lifespan than traditional bulbs, reducing the need for frequent replacements and associated energy usage.

Proper lighting controls are also essential for energy efficiency in lighting design. Lighting controls allow the lighting system to respond to changing conditions and user needs, resulting in reduced energy consumption. Examples of lighting controls include occupancy sensors, daylight sensors, and timers, which were discussed in detail in the previous questions. By using lighting controls, the lighting system can be turned off when not needed or dimmed when less light is required, resulting in significant energy savings.

Designing for daylighting can also improve energy efficiency in lighting design. Daylighting involves using natural light to provide illumination in a space, reducing the need for electric lighting. This can be achieved by using windows, skylights, or light shelves to allow natural light into the space. Proper daylighting design requires careful consideration of factors such as building orientation, window placement, and shading to ensure that the space receives adequate natural light without causing glare or unwanted heat gain.

Energy efficiency is a key consideration in sustainable lighting design. By using efficient light sources, proper lighting controls, and designing for daylighting, lighting systems can be designed to consume less energy while still meeting the lighting needs of the space [4, 5].

9.2 Use of Renewable Energy

The use of renewable energy sources in sustainable lighting design is becoming increasingly popular due to the need to reduce carbon emissions and reliance on non-renewable energy sources. Renewable energy sources such as solar, wind, and geothermal energy can be harnessed to power lighting systems and reduce the overall energy consumption of a building.

Solar power is a particularly promising source of renewable energy for lighting systems. Solar photovoltaic (PV) panels can be used to generate electricity from the sun's rays, which can be stored in batteries for use when sunlight is not available. This can be particularly useful for outdoor lighting systems, where there may be limited access to grid electricity.

Wind power can also be used to generate electricity for lighting systems through the use of small wind turbines. These turbines can be installed on the roof or in the vicinity of a building to generate electricity from the wind.

Geothermal energy is another renewable energy source that can be used for lighting systems. Geothermal heat pumps can be used to provide heating and cooling for a building, which can in turn reduce the need for artificial lighting by increasing natural daylighting and improving indoor air quality.

The use of renewable energy sources in lighting systems can have a significant impact on a building's carbon footprint and energy costs. However, it is important to consider the cost-effectiveness and feasibility of these solutions on a case-by-case basis [6, 7].

9.3 Daylight Harvesting

Daylight harvesting is the practice of using natural daylight to supplement or replace artificial lighting in a building. This technique involves the use of sensors, controls, and design strategies to optimize the use of natural light and reduce the need for electric lighting.

In sustainable lighting design, daylight harvesting is an important strategy for reducing energy consumption and improving the quality of light in indoor spaces. The benefits of daylight harvesting include lower energy costs, reduced greenhouse gas emissions, and improved visual comfort and well-being for building occupants [8].

There are several methods of daylight harvesting including the following.

9.3.1 Fenestration Design

The design of windows, skylights, and other openings aims to maximize natural light while minimizing glare and heat gain. Fenestration design is an essential aspect of daylight harvesting in sustainable lighting design. Fenestration refers to the design and arrangement of windows, skylights, and other openings in a building envelope to optimize natural light entry while minimizing heat gain or loss. Effective fenestration design can help reduce energy consumption by allowing the maximum amount of natural light to enter the building, reducing the need for artificial lighting and improving occupant comfort and productivity.

Some key design considerations for fenestration in daylight harvesting include window orientation, size, and glazing type. For example, windows facing north in the northern hemisphere receive the most consistent and diffuse light, while those facing south are subject to more direct sunlight and heat gain. Similarly, the size and placement of windows can be optimized to provide daylighting to interior spaces while avoiding glare and unwanted heat gain.

Different types of glazing can also affect the performance of fenestration systems in daylight harvesting. For example, low-emissivity (low-e) coatings and insulated glass units can reduce heat loss and gain, while high-transparency, spectrally selective glazing can maximize visible light transmission while reducing unwanted heat gain.

Effective fenestration design is an integral part of holistic daylight harvesting strategies that consider the overall building design, lighting system, and occupant needs. It requires a thorough understanding of the building's location, orientation, and climate, as well as the functional and aesthetic requirements of the interior spaces [8, 9].

9.3.2 Light Shelves

Horizontal surfaces reflect daylight onto the ceiling, distributing it evenly throughout the space. Light shelves are horizontal surfaces that reflect natural light deeper into a building's interior, increasing the effectiveness of daylight harvesting. Light shelves are commonly installed on the exterior of a building's facade and are typically made of reflective materials such as aluminum or stainless steel. The reflective surface helps to redirect and distribute natural light into a space, reducing the need for artificial lighting and saving energy [10, 11].

There are different types of light shelves, including top or bottom-mounted, fixed or adjustable, and horizontal or sloping. The type of light shelf chosen depends on factors such as the building's location, orientation, and window configuration.

However, the effectiveness of light shelves depends on the orientation of the building and the position of the sun throughout the day. Proper design and placement of light shelves are critical to maximize their effectiveness and achieve optimal daylighting levels [10, 11].

9.3.3 Light Tubes

Tubes or ducts that bring natural light into interior spaces from the roof or other external areas. Light tubes, also known as tubular skylights or sun tubes, are a type of daylighting system used in sustainable lighting design. Light tubes consist of a roof-mounted dome, which captures sunlight, and a reflective tube that directs the sunlight to a diffuser located in the ceiling of a building's interior. This allows natural light to enter the space and reduces the need for electric lighting during the day [12].

One of the advantages of light tubes is their ability to provide natural light to areas of a building that may not have access to windows or other forms of daylighting. They are particularly useful in spaces such as corridors, stairwells, and bathrooms. Light tubes are also relatively easy to install and maintain, and they can be used in both new and existing buildings.

However, there are some limitations to the use of light tubes. Their effectiveness depends on the amount of available sunlight, which can vary depending on the climate and time of day. Additionally, the quality and color of the light can differ from that of natural daylight, which can affect the overall aesthetic of a space.

Despite these limitations, light tubes are a popular choice for sustainable lighting design, particularly in commercial and institutional buildings. By incorporating natural light into the built environment, they can help reduce energy consumption and improve the health and well-being of occupants [12].

9.3.4 *Reflective Surfaces*

The use of reflective surfaces, such as light-colored walls, ceilings, and floors, helps disperse natural light throughout a space. Reflective surfaces play an important role in daylight harvesting by redirecting sunlight into interior spaces and maximizing the use of natural light. These surfaces can be in the form of mirrors, polished metals, or reflective films applied to surfaces such as walls, ceilings, or floors. By reflecting light into darker areas of a space, reflective surfaces can improve the overall lighting levels and reduce the need for electric lighting [13, 14].

The use of reflective surfaces in daylight harvesting has been shown to have significant benefits in terms of energy savings, occupant comfort, and productivity.

One example of the use of reflective surfaces in daylight harvesting is the Daylight Redirecting Film developed by 3M. This film is applied to windows and redirects up to 80% of incoming sunlight toward the ceiling, increasing natural light and reducing the need for electric lighting.

The use of reflective surfaces in daylight harvesting is an effective and sustainable way to maximize natural light and reduce energy consumption in buildings [13, 14].

9.3.5 *Daylight Sensors*

These are sensors that automatically adjust the electric lighting in response to changes in natural light levels, ensuring that artificial lighting is only used when needed. Daylight sensors are an important component of daylight harvesting systems, as they help regulate the amount of artificial lighting needed in a space based on available natural light. These sensors work by measuring the amount of light in a room and adjusting the artificial lighting accordingly to maintain a consistent level of illumination [15].

There are several types of daylight sensors:

- Photosensors: These sensors measure the amount of daylight entering a space and adjust the artificial lighting accordingly. They can be used in both indoor and outdoor applications.
- Task sensors: These sensors are used in spaces where specific tasks are performed, such as reading or computer work. They adjust the lighting in the task area based on the available natural light.
- Area sensors: These sensors measure the amount of daylight in a specific area and adjust the artificial lighting in that area accordingly.

Daylight sensors can be used in conjunction with other lighting control strategies, such as occupancy sensors and time-based controls, to create a more energy-efficient lighting system.

Daylight harvesting can be used in conjunction with other sustainable lighting strategies, such as LED lighting and lighting controls, to create a comprehensive lighting design that maximizes energy efficiency and minimizes environmental impact [15].

9.4 Light Pollution Reduction

Light pollution reduction is an important aspect of sustainable lighting design, as it seeks to minimize the negative impact of artificial lighting on the environment and wildlife. Here are some key points on this topic:

- Light pollution can have harmful effects on wildlife, including disrupting migration patterns and interfering with reproduction cycles. It can also have negative impacts on human health, including sleep disruption and increased risk of certain diseases.
- Sustainable lighting design seeks to minimize light pollution by using directional lighting, minimizing the amount of light that is directed upward or outward into the environment, and using shielding to direct light only where it is needed.
- Lighting designers can also use software to model the effects of proposed lighting designs on the environment and wildlife and adjust the design accordingly to minimize negative impacts.
- In addition to minimizing light pollution, sustainable lighting design can also incorporate measures to reduce energy consumption and greenhouse gas emissions, such as using energy-efficient lighting technologies and renewable energy sources [16].

9.5 Use of Environmentally Friendly Materials

In sustainable lighting design, the use of environmentally friendly materials is an important consideration. This involves selecting materials that are non-toxic, recyclable, and have low environmental impact [17].

The following are some examples of environmentally friendly materials used in lighting design:

- LED Lights: LED lights are energy-efficient and long-lasting, making them an ideal choice for sustainable lighting. They also contain no mercury or lead, reducing their environmental impact.
- Natural Materials: Natural materials like wood, bamboo, and stone are renewable and biodegradable. They can be used to create lighting fixtures that are both beautiful and sustainable.
- Recycled Materials: Recycled materials like glass, metal, and plastic can be used to create lighting fixtures that have a lower environmental impact. These materials reduce the need for new resources and help to divert waste from landfills.
- Low VOC Paints: Volatile organic compounds (VOCs) are chemicals found in many paints that can contribute to air pollution and indoor air quality problems. Low VOC paints are available that are healthier for people and the environment.
- Energy-Efficient Ballasts: Ballasts are used in fluorescent lighting to regulate the current. Energy-efficient ballasts can help reduce energy consumption and save money over time.

By using environmentally friendly materials in lighting design, we can reduce our impact on the environment and create a more sustainable future [17].

9.5.1 LED Lights

LED lights are a popular choice in sustainable lighting design due to their many environmental benefits. LEDs are energy-efficient, long-lasting, and contain fewer hazardous materials than traditional light sources. Here are some key points about the use of LED lights as an environmentally friendly material in lighting design:

- Energy efficiency: LEDs are much more energy-efficient than traditional incandescent bulbs, as they use up to 80% less energy to produce the same amount of light. This means that they reduce the amount of electricity needed to power lighting systems, resulting in lower energy bills and reduced greenhouse gas emissions.
- Long lifespan: LEDs have a much longer lifespan than traditional bulbs, lasting up to 25 times longer. This reduces the need for frequent replacements, which reduces waste and saves money over time.

- Lower environmental impact: LEDs contain fewer hazardous materials than traditional bulbs, such as mercury, lead, and other toxic metals. This makes them less harmful to the environment and easier to dispose of.
- Dimmable options: LED lights are compatible with dimming systems, which allows for better control of lighting levels and can reduce energy usage even further [18, 19].

9.5.2 Natural Materials

Natural materials, such as wood, cork, bamboo, and other plant-based materials, are becoming increasingly popular in lighting fixtures and design elements. These materials are renewable, biodegradable, and require less energy to produce than traditional materials like metal or plastic. In addition, they can help create a more visually appealing and calming environment.

One example of natural materials in lighting design is the use of bamboo as a lampshade material. Bamboo is a fast-growing, renewable resource that can be easily harvested without damaging the environment. It is also lightweight, durable, and has a unique texture and color that can add a natural and organic feel to any space [19].

Another example is the use of cork as a lighting fixture material. Cork is a natural, sustainable material that can be harvested from the bark of cork oak trees without harming the trees themselves. It is lightweight, fire-resistant, and has excellent sound-absorbing properties, making it an ideal material for both lighting fixtures and acoustic panels.

LED lights are also a popular choice for environmentally friendly lighting. They are highly energy-efficient, use less electricity than traditional lighting, and have a longer lifespan. This means that they require less maintenance and replacement, reducing the amount of waste generated from lighting systems.

The use of environmentally friendly materials in lighting design can help reduce the environmental impact of lighting systems and create a more sustainable and visually pleasing environment [19, 20].

9.5.3 Recycled Materials

The use of recycled materials in sustainable lighting design is an effective way to reduce waste and minimize the environmental impact of lighting systems. Recycled materials can be used in various components of lighting systems, including fixtures, lamps, and controls. For example, recycled glass can be used to make lamp shades, while recycled aluminum can be used in fixture bodies. Additionally, recycled plastics can be used in the production of lighting controls.

The use of recycled materials in lighting design also has economic benefits, as it can lower production costs and provide a more sustainable source of materials.

Here are some examples of the use of recycled materials in lighting design:

- The "Discocó" lamp by designer Christophe Mathieu for the Spanish lighting manufacturer Marset uses recycled polypropylene to create a series of circular discs that reflect and diffuse light.
- The "Recycled Tube Light" by Dutch designer Pieke Bergmans is made from discarded fluorescent tubes that have been cut, sanded, and fitted with energy-efficient LED bulbs.
- The "Oyster" LED pendant light by Australian designer Coco Reynolds is made from recycled PET plastic bottles that have been shaped into a series of oyster-like forms.

Overall, the use of recycled materials in lighting design is a sustainable and environmentally friendly approach that can help reduce waste and promote a more circular economy [20, 21].

9.5.4 Low VOC Paints

Low VOC (volatile organic compounds) paints are a type of environmentally friendly paint that emits lower levels of toxic chemicals compared to traditional paint. The use of low VOC paints in sustainable lighting design can help improve indoor air quality by reducing the levels of harmful pollutants in the air.

In addition to their environmental benefits, low VOC paints are also known for their durability, resistance to mold and mildew, and ability to dry quickly. They are available in a variety of colors and finishes and can be used in a wide range of applications, including walls, ceilings, and trim. There are several organizations that certify low VOC paints for use in sustainable building design, including the Leadership in Energy and Environmental Design (LEED) program and Green Seal. These certifications help ensure that the paint meets strict environmental and health standards.

The use of low VOC paints is an important aspect of sustainable lighting design, as it contributes to the overall health and well-being of building occupants while minimizing the impact on the environment [21, 22].

9.5.5 Energy-Efficient Ballasts

Energy-efficient ballasts are electronic devices that regulate the electrical current in lighting systems, particularly in fluorescent lamps, light-emitting diodes (LEDs), and other discharge lamps. They improve the efficiency of lighting systems and reduce energy consumption by controlling the flow of electricity, thereby preventing wastage and extending the lifespan of lighting systems. Energy-efficient ballasts are designed to meet various standards such as high-efficiency ballasts (HEB), low-frequency electronic ballasts (LFE), and programmed rapid-start ballasts (PSB). HEBs and LFEs are designed to minimize the energy consumed by the ballast, while PSBs are used in cold temperature applications and for rapid starting of lamps. Additionally, energy-efficient ballasts can also be paired with dimmers and sensors for advanced lighting control, further increasing energy efficiency.

The use of energy-efficient ballasts is a crucial aspect of sustainable lighting design as it helps minimize energy consumption and reduce carbon footprint. Furthermore, energy-efficient ballasts can save money on energy bills, reduce maintenance costs, and contribute to a healthier environment [23].

9.6 Long Lifespan

Long lifespan is a critical aspect of sustainable lighting design, as it contributes to reducing the environmental impact of lighting systems. The longer a lighting system lasts, the less energy and resources are required to replace it, resulting in less waste and pollution.

In sustainable lighting design, various strategies can be employed to extend the lifespan of lighting systems, including the use of high-quality components, proper maintenance, and designing for modularity and upgradability. By considering these factors, lighting designers can ensure that their systems last as long as possible, reducing the need for replacements and minimizing their environmental impact.

Moreover, long lifespan is closely related to the durability of lighting fixtures and their ability to withstand environmental conditions, such as heat, moisture, and physical impacts. Sustainable lighting designs aim to use fixtures made from high-quality, durable materials that can withstand harsh environmental conditions, reducing the need for frequent replacements.

In addition to reducing waste and pollution, long lifespan also results in cost savings over time, as fewer replacements are needed, and maintenance costs are minimized [1, 18].

9.7 Lighting Controls

Lighting controls play a crucial role in sustainable lighting design by enabling the optimization of lighting usage, reducing energy consumption, and enhancing occupant comfort and productivity. Here are some key aspects of lighting controls in sustainable lighting design:

- Dimming and Occupancy Sensors: The use of dimming controls and occupancy sensors can reduce energy consumption by automatically adjusting lighting levels based on occupancy and daylight levels.

- Daylight Harvesting: Daylight harvesting involves using daylight sensors to automatically adjust artificial lighting levels based on the available natural light, reducing the need for artificial lighting and improving energy efficiency.
- Time Scheduling: Scheduling controls can turn off or dim lights during periods of low occupancy, reducing energy consumption and extending lamp life.
- Networked Lighting Controls: Networked lighting controls allow for centralized management of lighting systems, enabling real-time monitoring, reporting, and remote control of lighting usage.
- Adaptive Lighting: Adaptive lighting systems use data from sensors and other sources to adjust lighting levels based on individual preferences and occupancy patterns, enhancing occupant comfort and energy efficiency.

By combining these approaches, lighting designers can create sustainable lighting systems that not only reduce energy consumption but also enhance occupant well-being and productivity [18].

9.7.1 Dimming and Occupancy Sensors

Dimming and occupancy sensors are important components of lighting control systems that can help increase energy efficiency and reduce electricity consumption. Dimming sensors allow for the adjustment of light output depending on the amount of natural light available or the time of day, while occupancy sensors detect the presence of people in a room and automatically turn lights on or off as needed.

Dimming and occupancy sensors can be used in a variety of settings, including offices, schools, hospitals, and residential buildings. Some lighting control systems even allow for the integration of both dimming and occupancy sensors for maximum energy efficiency [18, 20].

9.8 Human-Centric Lighting (HCL)

Human-centric lighting (HCL) is an approach to lighting design that aims to enhance human health, well-being, and productivity by considering the physiological and psychological effects of light on human beings. HCL systems take into account factors such as light intensity, color temperature, and spectral content to create lighting environments that align with the natural circadian rhythm of the human body. The concept of HCL is based on the idea that light is not just a functional requirement but also has a significant impact on human physiology, behavior, and psychology.

Research has shown that HCL systems can improve sleep quality, mood, cognitive performance, and overall well-being. HCL can also be used to support circadian rhythms in shift workers or in patients in healthcare settings. By using tunable white light or colored light, HCL systems can mimic natural daylight and create dynamic lighting scenes that change throughout the day [24].

One of the key aspects of HCL is the use of lighting controls that allow for fine-tuning of the lighting environment. This can include occupancy sensors, daylight sensors, and personal controls, such as dimmers or color temperature controls. By providing individual control over lighting, HCL systems can create personalized lighting environments that meet the needs of individual users. Designers interested in HCL must consider several factors, including the type of lighting system used, the color temperature and spectral content of the light, and the timing and duration of light exposure. They must also consider the specific needs of the occupants of a space, such as their age, activity levels, and visual impairments [25].

Some examples of HCL applications include healthcare facilities, schools, and workplaces. In healthcare settings, HCL systems can help patients recover faster by supporting their circadian rhythms and reducing stress. In schools, HCL systems can enhance learning outcomes by improving alertness and attention. In workplaces, HCL systems can boost productivity by reducing fatigue and improving mood.

There are some key features of human-centric lighting:

- Dynamic lighting: Human-centric lighting systems can automatically adjust the color and intensity of lighting throughout the day to mimic natural daylight patterns, which can help regulate circadian rhythms and improve sleep.
- Personalization: Human-centric lighting systems can be personalized to individual preferences and needs, allowing users to choose lighting settings that are most comfortable for them.
- Task-based lighting: Human-centric lighting systems can be designed to provide the right amount and type of light for specific tasks, such as reading or working on a computer, which can reduce eye strain and improve productivity.
- Flicker-free lighting: Human-centric lighting systems should be designed to minimize or eliminate flicker, which can cause headaches and eye strain.
- Glare reduction: Human-centric lighting systems should be designed to minimize glare, which can cause discomfort and visual impairment.
- Integration with other systems: Human-centric lighting systems can be integrated with other building systems, such as HVAC and security, to create a holistic building automation system.
- Energy efficiency: Human-centric lighting systems should be designed to be energy-efficient, using LED lighting and other efficient technologies to minimize energy use and reduce operating costs.
- User feedback: Human-centric lighting systems can incorporate user feedback to continuously improve and refine the lighting experience, ensuring that users are getting the most benefit from the system [24, 25].

9.8.1 Dynamic Lighting

Dynamic lighting systems are a type of human-centric lighting that are designed to adapt to the needs of individuals or groups of people in a given space. These systems use a combination of lighting controls, sensors, and algorithms to create lighting that mimics the natural changes in light throughout the day, providing the appropriate light intensity and color temperature for different activities and times of day [26].

Dynamic lighting systems have been shown to have a positive impact on human health and well-being, including improving sleep quality, reducing depression and anxiety, and increasing productivity and alertness. These systems can be particularly beneficial in spaces where people spend a significant amount of time, such as schools, hospitals, and office buildings.

There are a variety of different approaches to implementing dynamic lighting systems, including the following:

- Circadian rhythm lighting: This approach involves using lighting to simulate the natural progression of light throughout the day, with warmer, lower-intensity light in the morning and evening and cooler, higher-intensity light during the day.
- Task-specific lighting: This approach involves using lighting that is optimized for specific activities, such as reading or computer work, to enhance performance and reduce eye strain.
- Personalized lighting: This approach involves using sensors and algorithms to create individualized lighting profiles for each user, based on their personal preferences and needs.
- Responsive lighting: This approach involves using sensors to detect changes in the environment, such as the presence of people or changes in natural light, and adjusting the lighting accordingly.

Dynamic lighting systems can be complex to design and implement, requiring careful consideration of factors such as the needs of the occupants, the characteristics of the space, and the capabilities of the lighting system. However, with the right approach, these systems can have a significant positive impact on human health and well-being [26, 27].

9.8.2 Personalization

Personalization is a key feature of human-centric lighting, as it allows individuals to customize the lighting in their environment to meet their specific needs and preferences. With the increasing popularity of smart lighting systems, personalization has become easier and more accessible than ever before [28].

One way to personalize lighting is through the use of color temperature and intensity settings. For example, individuals may prefer warmer or cooler colors of light, or they may require brighter or dimmer lighting depending on the task or time of day. By allowing individuals to adjust these settings, they can create a lighting environment that is tailored to their needs.

Another way to personalize lighting is through the use of individual control systems. This allows each person to control the lighting in their own workspace or living area, without affecting others around them. Personalized control systems can be integrated into smart lighting systems or other types of lighting controls to allow individuals to adjust the lighting as needed. Personalization in human-centric lighting can have a positive impact on the health and well-being of individuals. For example, personalized lighting can help regulate the body's circadian rhythms, improve sleep quality, and reduce eyestrain and fatigue [28, 29].

9.8.3 Task-Based Lighting

Task-based lighting is an essential component of human-centric lighting design that seeks to optimize visual comfort and support the task performance of individuals. Task lighting refers to the use of targeted light sources to enhance visibility and reduce glare in areas where visual tasks are performed, such as reading, writing, or using a computer. By providing the right light levels and distribution, task-based lighting can improve visual acuity, reduce eye strain, and increase productivity.

One approach to task-based lighting is to incorporate adjustable or flexible lighting systems that allow users to tailor the lighting to their specific needs. For example, adjustable desk lamps can provide focused illumination for reading or writing tasks, while dimming controls can allow users to reduce glare and adjust the ambient light levels in the space. Other strategies include the use of specialized lighting fixtures, such as under-cabinet lighting in kitchens or task lighting in offices, to provide targeted illumination where it is most needed.

Incorporating task-based lighting into human-centric lighting design requires an understanding of the specific tasks and activities that will be performed in a given space, as well as the visual requirements of those tasks. It also requires consideration of the lighting quality and color rendering properties of the light sources used, as well as the overall visual aesthetics of the space [18].

9.8.4 Flicker-Free Lighting

Flicker-free lighting is an essential aspect of human-centric lighting design that ensures the lighting system's smooth operation without causing any discomfort to human beings. Flicker refers to the rapid and repeated changes in the light output of a light source that occur at a rate faster than the eye can perceive, typically around 50–60 Hz. This can cause discomfort, eye strain, headaches, and other health problems in some individuals, particularly those who are sensitive to light or have certain medical conditions [30].

To address this issue, human-centric lighting design incorporates flicker-free lighting technology. This technology ensures that the lighting system's light output is stable and free of flicker, providing a more comfortable and healthy lighting environment for occupants. Some common methods for achieving flicker-free lighting include selecting high-quality light sources, using electronic ballasts, and incorporating specialized lighting controls that help minimize flicker.

In addition to reducing discomfort and improving occupant health, flicker-free lighting can also enhance visual performance, reduce visual fatigue, and improve overall visual comfort. This is particularly important in environments where tasks requiring high visual acuity are performed, such as offices, classrooms, and healthcare facilities [18, 30].

9.8.5 Glare Reduction

Glare reduction is an essential aspect of human-centric lighting as it ensures the safety and comfort of individuals. Glare refers to the excessive brightness that causes discomfort and reduces visual performance. Inappropriate lighting designs can create excessive glare that may cause eyestrain, headaches, and even accidents. Therefore, human-centric lighting designs should incorporate glare reduction techniques to improve visual comfort and safety.

One of the main ways to reduce glare is through the use of anti-glare luminaires, which are designed to minimize the amount of light emitted at angles that could cause glare. Other effective techniques for glare reduction include the use of diffusers, reflectors, and louvers. These techniques help to soften and distribute the light, reducing the intensity and direction of the light beams. In addition to using anti-glare luminaires and glare reduction techniques, it is essential to consider the luminance levels in the lighting design. Luminance refers to the amount of light emitted by a surface, and it plays a crucial role in determining the glare levels. Therefore, human-centric lighting designs should maintain appropriate luminance levels to ensure optimal visual performance and reduce the risk of glare.

Overall, glare reduction is an important aspect of human-centric lighting that enhances visual comfort, safety, and productivity. By incorporating anti-glare luminaires, diffusers, reflectors, and appropriate luminance levels, human-centric lighting designs can minimize glare and optimize visual performance [29].

9.8.6 Integration with Other Systems

Human-centric lighting systems can be integrated with other building systems to enhance overall building performance and occupant comfort. Integration with other systems can provide benefits such as increased energy efficiency, better control over lighting systems, and improved indoor environmental quality. The following are some examples of how human-centric lighting can be integrated with other building systems:

- Building automation systems (BAS): Human-centric lighting can be integrated with a BAS to provide centralized control over all building systems. This allows for seamless control and monitoring of lighting, heating, ventilation, and air conditioning (HVAC), security, and other systems. Integration with a BAS also enables automated control of lighting systems based on occupancy, daylight levels, and other factors.

- Daylighting systems: Human-centric lighting can be integrated with daylighting systems to maximize the benefits of natural light. Daylight sensors can be used to adjust the intensity of artificial lighting in response to available daylight, providing consistent light levels throughout the day and reducing energy consumption.
- HVAC systems: Human-centric lighting can be integrated with HVAC systems to improve indoor environmental quality. For example, lighting can be used to regulate temperature and humidity levels by adjusting the intensity and color temperature of light sources.
- Occupancy sensors: Human-centric lighting can be integrated with occupancy sensors to provide adaptive lighting based on user behavior. For example, lighting levels can be adjusted based on the number of occupants in a room, their activity levels, and the time of day.
- Smart home systems: Human-centric lighting can be integrated with smart home systems, allowing occupants to control lighting from their mobile devices. Smart home systems can also be used to automate lighting based on user preferences and behavior.

Integration with other building systems is essential to achieve the full potential of human-centric lighting. By working together, these systems can create a comfortable and productive indoor environment while reducing energy consumption and operating costs [18, 29].

9.8.7 Energy Efficiency

Human-centric lighting not only focuses on providing appropriate light levels, but it also aims to achieve energy efficiency. Several lighting technologies and control strategies can be implemented to ensure energy efficiency while also providing quality lighting for the users. Here are some ways to achieve energy efficiency in human-centric lighting:

- LED Lighting: LED lights are the most efficient lighting source available, consuming up to 90% less energy than traditional incandescent bulbs. LEDs also have a longer lifespan, reducing the need for frequent replacements, which can save energy and maintenance costs.
- Lighting Controls: Lighting controls such as occupancy sensors, daylight sensors, and dimming controls can help reduce energy consumption by turning off or dimming lights in unoccupied or well-lit areas. This not only saves energy but also improves user comfort by reducing glare and maintaining consistent light levels.

- Task-Based Lighting: Providing lighting based on the task being performed can help reduce energy consumption. By illuminating the necessary areas, energy is saved by not lighting the entire space. Task-based lighting also provides better visual comfort, which can lead to improved productivity and well-being.
- Automated Lighting: Automated lighting systems can help reduce energy consumption by adjusting the lighting levels and color temperature based on occupancy, time of day, and other factors. These systems can also be integrated with other building systems to maximize energy efficiency [31].

By incorporating energy-efficient lighting technologies and control strategies, human-centric lighting can provide quality lighting while reducing energy consumption and promoting user feedback:

User feedback is an important aspect of human-centric lighting design. In order to create lighting that meets the needs of individuals and provides the desired benefits, it is important to gather feedback from users and incorporate that feedback into the design.

One way to gather feedback is through surveys and interviews with building occupants. This can provide valuable insight into how people use the space and what their lighting preferences are. Another approach is to use sensors to monitor the usage of the lighting system and gather data on how it's being used.

Once feedback is collected, designers can use that information to adjust the lighting system to better meet the needs of users. This might involve changing the color temperature, adjusting the intensity or timing of the lighting, or incorporating different types of fixtures or controls.

In addition to improving the lighting system itself, user feedback can also help to educate occupants about the benefits of human-centric lighting and how to use the system to achieve their desired outcomes [29, 31, 32].

9.9 Maintenance and Serviceability

Maintenance and serviceability are critical aspects of sustainable lighting design. Efficient maintenance and serviceability ensure that lighting systems continue to function optimally over an extended period, reducing the need for frequent replacement and minimizing waste. Proper maintenance can also contribute to energy savings by ensuring that lighting systems operate at peak efficiency.

To achieve efficient maintenance and serviceability in sustainable lighting design, several strategies can be employed. First, lighting systems should be designed for easy access and repair, minimizing the need for costly and time-consuming maintenance procedures. Second, regular maintenance schedules should be implemented, including routine cleaning and replacement of components such as lamps, ballasts, and controls. Third, lighting systems should be equipped with remote monitoring and control capabilities, allowing maintenance personnel to identify and address issues promptly [33].

Incorporating maintenance and serviceability considerations into lighting design requires a comprehensive approach that considers the entire lifecycle of the system. This approach ensures that systems are designed for optimal performance, longevity, and energy efficiency, reducing the environmental impact of lighting systems and contributing to sustainable lighting design [34].

There are some of the main features of maintenance and serviceability in sustainable lighting design:

- Easy access to lighting fixtures and controls for maintenance and repair
- Use of modular lighting systems that allow for easy replacement of individual components
- Integration of lighting controls that allow for remote monitoring and adjustment of lighting levels
- Use of durable and long-lasting materials to minimize the need for replacement
- Regular cleaning and maintenance to ensure optimal performance and energy efficiency
- Implementation of a comprehensive maintenance plan that includes routine inspections and cleaning, as well as scheduled replacement of components nearing the end of their useful life
- Use of energy-efficient lighting sources and controls that minimize the need for maintenance and replacement.

These features help ensure that sustainable lighting systems are designed for long-term use and ease of maintenance, while also maximizing energy efficiency and minimizing environmental impact [33, 34].

References

1. Lechner, N. (2014). *Heating, cooling, lighting: Sustainable design methods for architects*. Wiley.
2. Balocco, C., & Volante, G. (2019). A method for sustainable lighting, preventive conservation, energy design and technology—Lighting a historical church converted into a university library. *Sustainability, 11*(11), 3145.
3. Tregenza, P., & Loe, D. (2013). *The design of lighting*. Routledge.
4. Jensen, C. L. (2017). Understanding energy efficient lighting as an outcome of dynamics of social practices. *Journal of Cleaner Production, 165*, 1097–1106.
5. Roufechaei, K. M., Bakar, A. H. A., & Tabassi, A. A. (2014). Energy-efficient design for sustainable housing development. *Journal of Cleaner Production, 65*, 380–388.
6. Martin, G. (2003). Renewable energy gets the "green" light in Chicago. *IEEE Power and Energy Magazine, 1*(6), 34–39.
7. Shi, L., & Chew, M. Y. L. (2012). A review on sustainable design of renewable energy systems. *Renewable and Sustainable Energy Reviews, 16*(1), 192–207.
8. Behera, A. R., Siddhartha, & Devi, J. (2018). Daylight harvesting concepts and its application: A review. In *Advances in smart grid and renewable energy: Proceedings of ETAEERE-2016* (pp. 65–81). Springer.
9. Wu, G. (2012). *Studies in preliminary design of fenestration: Balancing daylight harvesting and energy consumption* (Doctoral dissertation, University of Southern California).

10. Littlefair, P. J. (1995). Light shelves: Computer assessment of daylighting performance. *International Journal of Lighting Research and Technology, 27*(2), 79–91.
11. Kontadakis, A., Tsangrassoulis, A., Doulos, L., & Zerefos, S. (2017). A review of light shelf designs for daylit environments. *Sustainability, 10*(1), 71.
12. Laouadi, A. (2005). Design insights on tubular skylights. *Lighting Journal: Art and Science for International Designers, 25*, 38–41.
13. Katunský, D., Dolníková, E., Dolník, B., & Krajníková, K. (2022). Influence of light reflection from the wall and ceiling due to color changes in the indoor environment of the selected hall. *Applied Sciences, 12*(10), 5154.
14. Synnefa, A., & Santamouris, M. (2013). White or light colored cool roofing materials. *Advances in the Development of Cool Materials for the Built Environment, 2*, 33–71.
15. Gentile, N., Dubois, M. C., & Laike, T. (2015, June). Daylight harvesting control systems design recommendations based on a literature review. In *2015 IEEE 15th international conference on environment and electrical engineering (EEEIC)* (pp. 632–637). IEEE.
16. Gaston, K. J., Davies, T. W., Bennie, J., & Hopkins, J. (2012). Reducing the ecological consequences of night-time light pollution: Options and developments. *Journal of Applied Ecology, 49*(6), 1256–1266.
17. Casamayor Alarco, J. L., & Su, D. (2010). Sustainable lighting product design: A new approach and an industrial case study. In *Proceedings of the LeNS conference* (Vol. 2, pp. 1148–1162). Greenleaf Publishing Limited.
18. Fahima Zabeen, Z. (2016). *Sustainable lighting* (Doctoral dissertation).
19. Jägerbrand, A. K. (2015). New framework of sustainable indicators for outdoor LED (light emitting diodes) lighting and SSL (solid state lighting). *Sustainability, 7*(1), 1028–1063.
20. Cenci, M. P., Dal Berto, F. C., Schneider, E. L., & Veit, H. M. (2020). Assessment of LED lamps components and materials for a recycling perspective. *Waste Management, 107*, 285–293.
21. Casamayor, J. L., & Su, D. (2013). Integration of eco-design tools into the development of eco-lighting products. *Journal of Cleaner Production, 47*, 32–42.
22. Lam, P. T., Chan, E. H., Chau, C. K., & Poon, C. S. (2011). A sustainable framework of "green" specification for construction in Hong Kong. *Journal of Facilities Management, 9*(1), 16–33.
23. Li, D. H., Cheung, K. L., Wong, S. L., & Lam, T. N. (2010). An analysis of energy-efficient light fittings and lighting controls. *Applied Energy, 87*(2), 558–567.
24. Di Nicolantonio, M., Rossi, E., Deli, A., & Marano, A. (2020). The human centric lighting approach for the design of Age-Friendly products. *Theoretical Issues in Ergonomics Science, 21*(6), 753–772.
25. Malmberg, J. (2017). *Is human centric lighting really human centric in a classroom?: A holistic evaluation of a HCL-system in a classroom*. KTH Royal Institute of Technology, School of Architecture and the Built Environment.
26. Choi, K., & Suk, H. J. (2016). Dynamic lighting system for the learning environment: Performance of elementary students. *Optics Express, 24*(10), A907–A916.
27. Mott, M. S., Robinson, D. H., Walden, A., Burnette, J., & Rutherford, A. S. (2012). Illuminating the effects of dynamic lighting on student learning. *SAGE Open, 2*(2), 2158244012445585.
28. Papatsimpa, C., Bonarius, J. H., & Linnartz, J. P. M. (2020). Human centric iot lighting control based on personalized biological clock estimations. In *2020 IEEE 6th world forum on internet of things (WF-IoT)* (pp. 1–6). IEEE.
29. Walerczyk, S., Hclpc, C. L. E. P., & Wizards, L. L. (2012). Human centric lighting. *Architectural SSL, 23*, 20–22.
30. Burgardt, I., Agostini Junior, E., Illa Font, C. H., & Nascimento, C. B. (2016). Dimmable flicker-free power LEDs lighting system based on a SEPIC rectifier using a regenerative snubber. *IET Power Electronics, 9*(5), 891–899.
31. Houser, K. W. (2018). Human centric lighting and semantic drift. *LEUKOS, 14*(4), 213–214.

32. Madias, E. N. D., Christodoulou, K., Androvitsaneas, V. P., Skalkou, A., Sotiropoulou, S., Zervas, E., & Doulos, L. T. (2023). The effect of artificial lighting on both biophilic and human-centric design. *Journal of Building Engineering, 76*, 107292.
33. Doultsinou, A., Roy, R., Baxter, D., Gao, J., & Mann, A. (2009). Developing a service knowledge reuse framework for engineering design. *Journal of Engineering Design, 20*(4), 389–411.
34. Asmone, A. S., Conejos, S., & Chew, M. Y. (2019). Green maintainability performance indicators for highly sustainable and maintainable buildings. *Building and Environment, 163*, 106315.

Index

G. Ozenen, *Architectural Interior Lighting*,
https://doi.org/10.1007/978-3-031-49695-0

MIX
Papier aus verantwortungsvollen Quellen
Paper from responsible sources
FSC® C105338

If you have any concerns about our products,
you can contact us on
ProductSafety@springernature.com

In case Publisher is established outside the EU,
the EU authorized representative is:
Springer Nature Customer Service Center GmbH
Europaplatz 3, 69115 Heidelberg, Germany

Printed by Libri Plureos GmbH
in Hamburg, Germany